CLASSIC GEOLOGY

Italian volcanoes

Italian volcanoes

Chris Kilburn & Bill McGuire
University College London

TERRA

First published in 2001 by Terra Publishing. Reprinted 2011.

Terra Publishing (Dunedin Academic Press Ltd)
Hudson House
8 Albany Street
Edinburgh EH1 3QB
Scotland
www.dunedinacademicpress.co.uk

ISBN 978-1-903544-04-4

British Library Cataloguing-in-Publication Data
A CIP record for this book is available from the British Library

Library of Congress Cataloging-in-Publication Data are available

Typeset in Palatino and Helvetica
Printed and bound by CPI Group (UK) Ltd, Croydon, CR0 4YY

Contents

v

Preface

This book began as a guide to Italy's active volcanoes and finished as a guide and a practical introduction to volcanology. The evolution was inevitable. Across the few hundred kilometres that connect Sicily's volcanoes to their Neapolitan cousins, virtually all major volcanic products can be discovered with ease. Add to that the charm and excitement of Mediterranean life and it is clear that Italy is among the best and most enjoyable countries to learn about volcanoes and their impact on society.

Chapter 1 introduces Italian volcanism, from its relations with tectonic setting to styles of eruption. It also contains guidelines for working safely on active volcanoes and these should be read *before going into the field*. The remaining chapters are dedicated to four key volcanic districts: Somma–Vesuvius, Campi Flegrei, the Aeolian Islands and Etna. Each chapter discusses the evolution of the district and its importance to understanding volcanic processes, including accounts of classic eruptions. Recommended itineraries are then described (at least two per chapter), followed by a selection of other sites that can be added as desired. To maintain an informal style, relevant sources have not been referenced as for a specialist journal or standard work of reference; instead, we have gathered by chapter lists of key publications and Internet sites. We hope our friends and colleagues will indulge us in taking this mild professional liberty.

As a final note of caution, Italy's volcanoes can be dangerously seductive. We hope this book will add to the pleasure of offering no resistance.

Acknowledgements

The guide would not have been possible without the support of several friends and colleagues, in particular: Peter Cattermole, whose advice greatly improved our knowledge of the Aeolian Islands, Berthold Bauer (University of Vienna), Franco Mantovani (University of Ferrara) and participants of the 2000 Austro-Italian field course to southern Italy; Peppe Rolandi at the University of Naples; Maurizio Fraissinet, Carlo Bifulco, Emilio Musella, Stefano Carlino and Gino Menegazzi from the Vesuvius National Park, and Alwyn Scarth, whose comments improved an early draft of the guide. As always, special mention must also be made of the patience of the publishers.

Chapter 1

Volcanism in Italy

Italy is volcanically the busiest nation in continental Europe. At least 30 volcanic centres have erupted in the past five million years, of which a dozen have had outbursts within the past 10 000 years, seven are classified as still active (Somma–Vesuvius, Campi Flegrei, Etna, Stromboli, Vulcano, Lipari and Pantelleria), and two, Etna and Stromboli, are in virtually continuous eruption.

The central role of the Mediterranean region in the rise of European culture has resulted in Italy's volcanoes being also the best studied in the world. Etna, in particular, has an unprecedented historical record dating back to at least the second century BC, and observations at Somma–Vesuvius describe the volcano's activity since it destroyed Pompeii and Herculaneum in AD 79. For more than 2000 years, sailors have used Stromboli's eruptive plume – a cloudy column by day and orange glow by night – as a navigational beacon; at nearby Lipari, settlers have traded in the island's fine obsidian since prehistoric times.

In the eighteenth and nineteenth centuries, Naples and Somma–Vesuvius were obligatory stops on the European Grand Tour, a cultural journey undertaken by the aristocracy and its protégés. At the time, Naples was a royal capital (of the Kingdom of the Two Sicilies, incorporating southern Italy and Sicily) and Somma–Vesuvius was in persistent eruption. Add the fascination of Roman ruins being excavated at Pompeii and Ercolano and it is easy to understand why Somma–Vesuvius soon became the most famous volcano on the planet.

As well as popularizing Somma–Vesuvius and, to a lesser extent, Etna in Sicily, the Grand Tour served to stimulate understanding about how volcanoes work. Indeed, the heady mix of local observations with the new geological thinking of the 1800s ignited a chain reaction that was to lay the foundations of modern volcanology. A tour of Italy's volcanoes will thus not only reveal the extraordinary range of volcanic activity but it will also trace a path through the cradle of volcanology.

Italy's volcanoes and tectonics

Italian volcanism is one result of the collision between the Eurasian and African tectonic plates. The complex nature of the collision is masked by the apparently simple arrangement of Italy's volcanoes along a front extending from Lardarello and Monte Amiata in Tuscany, past the Campanian volcanoes and Etna, to the tiny island of Pantelleria, two thirds of the way between Sicily and Tunisia (Fig. 1.1). In detail, however, this front consists of interacting segments that can be distinguished by the age, clustering and chemistry of its component volcanoes. Attempts to explain the interaction have inspired several imaginative models for the tectonics of the Italian region, and it is only during the past 10–15 years that the first signs of a consensus have finally emerged.

Geographically, Italy's volcanoes can be gathered into three main regions (Fig. 1.1): the Roman province extending from Monte Amiata to Colli Albani (Alban Hills); the Campanian province, which, embracing Roccamonfina, Campi Flegrei and Vesuvius, also lies along a crude east–west alignment from Vulture to Ponza; and the Aeolian Islands and Etna.

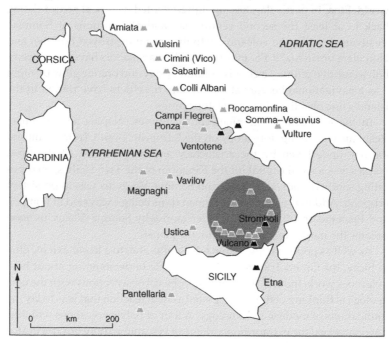

Figure 1.1 Quaternary volcanoes in Italy. The black volcanoes are those described in the guide. The Roman Province runs from Amiata to the Colli Albani (Alban Hills). The Campanian volcanoes (Roman Comagmatic Province) incorporate Roccamonfina, Campi Flegrei and Somma–Vesuvius. The Aeolian Islands are in the shaded circle north of Sicily.

These are supplemented by scattered offshore volcanoes, including Pantelleria, west-southwest of Sicily.

The principal geographical regions can in turn be combined into dynamic units according to their relative ages (Table 1.1). The Roman and Campanian volcanoes can be envisaged as ends of a unit along which the onset of volcanism has migrated southwards in time, from about 1.0–1.3 million years ago in the north to 0.3–0.5 million years in the Neapolitan district. Similarly, Ponza, Ventotene and Vulture form a trend, with volcanism migrating eastwards from 4.0–5.0 million years to 0.8–1.0 million years, and Etna and the Aeolian Islands are the active southeastern limb of a trend that began at the undersea volcanoes Magnaghi and Vavilov 3.5–7.5 million years ago (Table 1.1).

The growth and location of the dynamic units reflect the anticlockwise rotation of the region behind the Afro–Eurasian collision zone. Twisting stretched the continental crust to breaking point about 7.0–8.0 million years ago. Huge volumes of basalt escaped from the mantle to coat the thinning and sinking crust and to provide the nucleus of the present Tyrrhenian Basin. The reduced crustal thickness allowed the underlying mantle to rise to shallower depths, and the combination of general mantle uplift and eastward crustal motion produced a crude three-way zone of preferred crustal failure (roughly to the north, east and southwest) centred about 100 km west of the modern Campanian coast (Fig. 1.2).

The triple junction has since been used to concentrate regional spreading along two directions, separated by the east fracture zone (Fig. 1.2). Thus, the northern half of Italy has continued mainly to rotate anticlockwise, eruptions along the north fracture zone producing the Roman volcanic province and Roccamonfina. Across southern Italy and Sicily, rotation has been distorted by a strong southeastward stretching that has fragmented the original southwest fracture zone and favoured the migration of volcanism towards Sicily. Along the east fracture zone itself, magmatism has produced the volcanoes of Ponza, Ventotene and Vulture.

A striking feature of the onset times of volcanism (Table 1.1) is that, following very early activity at Ponza and Magnaghi–Vavilov (3.5 and 7.5 million years ago), a main phase of eruptions appears to have begun along *all* the fracture zones about 1 million years ago, with a second burst in the Neapolitan district and in Sicily about 500 000 years later. The former phase may reflect the crossing of a critical condition in the mantle or crust (such as large-scale melting in the mantle or deep-seated crustal failure). The later phase involved volcanoes along the borders of the southeastward-moving crust and may have been related to accelerated spreading and crustal tearing.

As well as controlling the locations of Italy's volcanoes, crustal

3

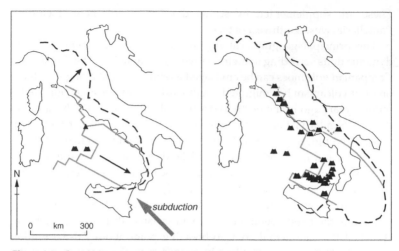

Figure 1.2 Crustal movement in the Italian region from 5 million years ago (left) to the present (right). The main collision front (dashed) has rotated anticlockwise and extended to the southeast. Crustal failure (grey lines) around a triple junction (triangle, left) has allowed different preferred movements between the northern and southern Italy (small arrows, left). Volcanoes (filled circles), identified in Figure 1.1, lie along migrating fault systems. Northwest subduction of the African plate (large arrow, left) continues to feed magma to the Aeolian Islands. Fault segments (now apparently inactive volcanically) link Vulture to the Campanian volcanoes (grey dotted lines, right). See Turco & Zuppette (1998).

Table 1.1 Approximate onset times of volcanism in Italy.

Start date (million years)	≤0.5	>0.5–1.0	>1.0–1.5	>1.5–2.0	>2.0–2.5	>2.5
Roman volcanoes			1.0–1.3			
Roccamonfina		0.7–1.0				
Neapolitan volcanoes	0.3–0.5					
Ponza						4.0–5.0
Ventotene			1.0–1.5			
Vulture		0.8–1.0				
Magnaghi & Vavilov						3.5–7.5
Marsili					~2.0	
Aeolian Islands			>1.3			
Etna	0.5					
Pantelleria	>0.3					

thinning, twisting and collision have produced a range of local tectonic settings for recent activity. The southeastward stretching has forced subduction of the African plate beneath Calabria, at the toe of Italy, inducing mantle melting and formation of the Aeolian island arc. Fifty kilometres south, local tension along the edge of the southeast unit has allowed magma to escape beneath Etna. Rotation of the northern unit, meanwhile, has triggered major block faulting and continued activity from the

Neapolitan volcanoes, Somma–Vesuvius and Campi Flegrei. Such a range of tectonic environments is unusual over distances as small as 500 km and a full interpretation remains the goal of current research.

The nature of Italian magmas

In common with most magmatic suites across the globe, the products of Italian volcanism consist mainly (about 99 per cent by weight) of eight elements: silicon, aluminium, magnesium, iron, calcium, sodium, potassium and oxygen. Of these, silicon and oxygen account for about one-half or more (Table 1.2) and, as magma solidifies, they form a series of silicate structures within which the remaining elements reside. The first silicates to form during cooling are those rich in the refractory elements magnesium, iron and calcium, and only later do additional silicates appear containing the geochemically similar alkali elements, sodium and potassium. As crystallization proceeds, therefore, the liquid fraction of a magma becomes progressively enriched in potassium and sodium, following a trend that depends on the starting composition of the molten magma and on the pressures and temperatures that the magma encounters during ascent.

The magma crystallization path is traced on a total alkali-silica (TAS) diagram (Fig. 1.3), SiO_2 being ever present in silicate minerals, and the alkali oxides (Na_2O+K_2O) providing a crude measure of the amount of crystallization that has occurred. By superimposing the compositional fields of common igneous rocks, the TAS diagram yields a simple means of representing evolutionary trends from one rock type to another (Fig. 1.3).

Table 1.2 Major-element compositions (per cent by weight) of selected Italian volcanic rocks.

Oxide	Hawaiite Etna	K-phonolitic tephrite Vesuvius	K-mugearite Somma	K-trachy-andesite Stromboli	K-phonolite Campi Flegrei	K-trachyte Somma	K-dacite Vulcano	K-rhyolite Vulcano
SiO_2	48.78	47.50	51.07	56.78	58.56	60.64	68.8	73.0
TiO_2	1.59	1.02	0.86	0.82	0.75	0.37	0.1	0.1
Al_2O_3	17.95	17.40	18.53	17.50	19.84	17.45	13.8	13.6
Fe_2O_3	5.08	3.21	4.66	2.62	2.56	1.52	1.7	1.2
FeO	4.52	5.10	2.76	3.94	0.80	1.72	1.4	1.1
MnO	0.16	0.16	0.12	0.16	0.20	0.14	0.1	0.1
MgO	5.54	4.12	4.75	2.87	0.24	0.34	0.6	0.2
CaO	9.65	9.93	7.58	6.73	1.71	2.78	1.7	1.0
Na_2O	3.90	2.59	2.41	3.63	6.61	3.64	3.5	4.5
K_2O	1.37	6.95	6.23	3.56	6.73	7.80	4.9	5.2

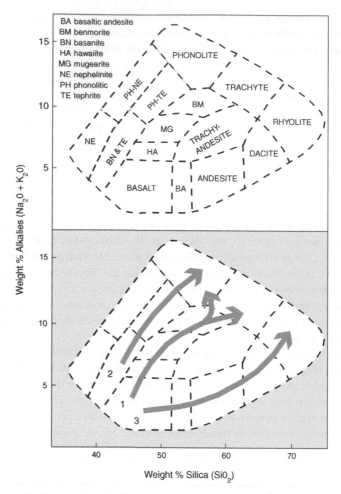

Figure 1.3 The standard classification (top) of volcanic rocks according to their total alkali (Na$_2$O+K$_2$O) and silica contents. Trachybasalts are divided into the three fields for hawaiite, mugearite and benmoreite. See Table 1.3 for some alternative nomenclature for potassic compositions. Common evolutionary trends (bottom) among Italian magmas are; (1) K-basalt to K-trachyte or K-phonolite, (2) K-tephrite to K-phonolite, and (3) K-basalt to K-rhyolite.

In the majority of magmas, the weight ratio K$_2$O/Na$_2$O in the liquid fraction is less than about 60 per cent. Only in a minority of cases are magmas relatively potassium enriched, these cases including the Roman and Campanian volcanoes and some of the Aeolian Islands. Since potassic suites are comparatively rare, their rock types have been given a confusing collection of exotic names, with the same names sometimes being used to define fields with different chemical limits. Some of the more frequently

Table 1.3 Common names for potassium-rich volcanic rocks.

Standard rock	Potassic equivalent
Basalt	Shoshonite
Tephrite	Leucite tephrite
Basanite	Leucite basanite
Trachybasalt (hawaiite, mugearite, benmoreite)	Leucite-trachybasalt; latite*
Trachyandesite	Latite
Rhyolite	K-rhyolite
Trachyte	K-trachyte
Phonolite	Leucitophyre

*Italian authors may extend the limits of the latite field to include some potassic trachybasalts.

used names are listed in Table 1.3. Except when common local usage demands otherwise, this guide favours simplicity over purist nomenclature and describes the magmas as the potassic versions of their standard equivalents; nor do we distinguish between potassic ($K_2O/Na_2O > 0.6$) and ultrapotassic ($K_2O/Na_2O > 1.6$) rocks.

Potassic volcanism is significant because its rarity suggests magma formation under unusual conditions. Current models favour re-melted continental crust as a ready source of additional potassium, K-enrichment becoming anomalous when the ratio of crustal to mantle melt exceeds a critical value. Indeed, tectonic settings noted for K-rich magmas are continental rift zones (where faulting and foundering promote crustal melting) and subduction zones in which the descending crustal slab contains some continental material (even after subduction has ceased, sections of subducted crust may still remain to be melted).

Given that opening of the Tyrrhenian Basin has involved the thinning and faulting of continental crust for 7–8 million years, it is not surprising that most of Italy's recent magmas have belonged to potassic suites (Fig. 1.3), notably the trends (1) K-basalt, K-trachybasalt, K-trachyte, and/or K-phonolite (e.g. Somma–Vesuvius and Campi Flegrei), (2) K-tephrite (leucite-tephrite), K-phonolitic tephrite, K-phonolite (e.g. Somma–Vesuvius), and (3) K-basalt, K-andesite, K-dacite, K-rhyolite, (e.g. Vulcano and modern Stromboli). Non-potassic trends also prevail at Etna (basalt, trachybasalt, trachyte) and were shown by some of Stromboli's prehistoric magmas (basalt, andesite, dacite, rhyolite).

Styles of eruption

Italy's active volcanoes are a living museum of volcanology. Their products embrace every major style of eruption, from gentle effusions of lava, through Strombolian, Vulcanian and hydromagmatic explosivity, to violent Plinian outbursts and caldera collapse. Lava effusion and Strom-

7

persistently active vents on Etna and Stromboli. However, the remaining styles have all occurred in historical time and can be expected again in the future.

How volcanoes erupt depends on their explosive potential near the surface and this, in turn, is controlled by the fluidity of magma and its concentration of bubbles (Fig. 1.4). Bubbles contain the gases emitted during eruptions. For the most part, they consist of water vapour and first appear at depths of 0.3–5 km, roughly corresponding to H_2O contents initially dissolved in deep magma from less than 1 per cent to about 3–4 per cent by weight. The dependence on depth is explained by pressure exerting a key control on when bubbles can form: just as opening a bottle of fizzy water lets bubbles form as pressure is released, so magmatic bubbles appear as the pressure exerted by the weight of overlying rock decreases while magma ascends towards the surface.

The more fluid a magma, the easier it is for a bubble to migrate upwards under buoyancy and to escape before the magma reaches the surface. This is especially the case if the magma is also rising slowly or has been temporarily stalled at shallow depth. The result is magma emerging as a lava flow (Fig. 1.5), containing perhaps 10–20 per cent by volume of bubbles normally a centimetre or less across. As magma resistance increases and its ascent velocity decreases, the enclosed bubbles are less able to escape before eruption. Pressures increase in the trapped bubbles and, when they have become high enough, they provide the conditions necessary for explosive activity.

The trapped bubbles can either grow independently as a froth or

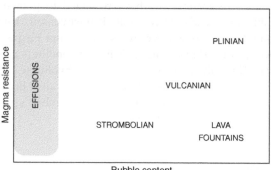

Figure 1.4 Variation of eruption style with magma resistance and bubble content. Magma generally becomes more resistant and bubble-rich as it evolves compositionally. Thus, although gas-poor fractions of all magmas can effuse lava flows and domes, Strombolian eruptions and lava fountains are commonly (but not exclusively) associated with basaltic, trachybasaltic and trachyandesitic magmas, Vulcanian eruptions with andesitic and trachytic magmas, and Plinian eruptions with andesitic, dacitic, trachytic and phonolitic magmas. Explosive rhyolitic eruptions have not been observed in historical times, but are assumed to have Plinian characteristics.

Figure 1.5 Aa lavas destroying the Ristorante Corsaro on Etna in 1983. The building was eventually buried completely. The new Hotel Corsaro has been built on top of the flows, directly above its predecessor. Note the rubbly nature of the aa surface.

coalesce into larger bubbles. When frothing dominates, magma erupts as a column of gas containing the molten fragments of burst bubble walls. Emerging as a jet, the column soon loses its larger and heavier fragments as they rain down close to the vent. Smaller and lighter fragments continue upwards, heating the air around them. As the initial force from the jet wanes (at heights from hundreds of metres to a few kilometres), the heated air (less dense than the cold air around it) carries the lighter fragments aloft under buoyancy. The result (Fig. 1.6) is a billowing eruption column that may rise 20 km or more through a thinning atmosphere until it is forced to spread laterally (at a level where the volcanic mixture and surrounding air share the same density). Fine particles (ash) from these clouds can travel hundreds of kilometres before settling back to Earth, the smallest even encompassing the globe if carried into the stratosphere.

The earliest written European account of a billowing eruption column is Pliny's description of the AD 79 outburst of Somma–Vesuvius. Accordingly, the parent events are known as Plinian eruptions. As a Plinian column rises, its outer layers continue to mix with the cold surrounding air and may eventually become too heavy for buoyant ascent. The heavy layers then collapse and, crashing from heights of several kilometres, send clouds of gas and ash racing over the ground at hurricane velocities. Such pyroclastic flows (formerly nuées ardentes or glowing clouds) have been among the greatest killers in volcanic eruptions (Fig. 1.7).

9

Figure 1.6 Vivid impressions of the sub-Plinian eruption of Somma–Vesuvius in 1779. The night view (above) clearly distinguishes the jet (lava fountain) from the buoyant ash cloud, whose billowing nature is seen more clearly by day (opposite). Note also the lightning induced by electrostatic charges in the ash cloud.

Figure 1.7 A modest pyroclastic flow racing down slope at velocities of 60–70 km an hour. This example from Montserrat in the Caribbean was triggered by the collapse and disintegration of a lava dome. Flows of similar appearance are produced by the collapse of Plinian eruption columns. In AD 79, a series of column-collapse pyroclastic flows from Somma buried Herculaneum, their high temperatures (~600°C or more) having previously killed evacuees waiting for boats along the shore.

Figure 1.8 Lava fountaining from Etna's Northeast Crater in November 1986. At their peak, the fountains reached heights of several hundred metres. Stray bombs landed 1–2 km from the vent. The change from modest explosivity to fountaining occurred in less than 30 minutes, illustrating the rapidity with which an eruption can become extremely dangerous for nearby onlookers. (See also Fig. 1.10.)

among the greatest killers in volcanic eruptions (Fig. 1.7).

When the magma is broken into fragments too large to be buoyed upwards, the eruptive jet develops as a lava fountain (Fig. 1.8). If in addition the rate of magma ascent has been just slow enough to allow bubbles to coalesce, the eruption may occur as a series of discrete explosions at almost regular intervals. In this case, each large explosion reflects the bursting of a giant bubble (large enough to fill a vent), the broken skin of which is hurled away as a collection of fragments sometimes several metres across (Fig. 1.9). Typical on Stromboli, this behaviour has been formally classified as Strombolian activity.

Intermittent explosions also occur among very resistant magmas in which trapped bubbles gather and expand below a thick magmatic cap, similar to the cork in a bottle. Weighed down by the cap, the bubbles develop pressures greater than those for Strombolian eruptions, so they trigger more violent explosions that, on occasion, can shatter the volcanic edifice. Since Giuseppe Mercalli's classic description of this behaviour during Vulcano's 1888–90 eruption (Ch. 4), such eruptions have been

13

Figure 1.9 Strombolian activity from Etna's Southeast Crater in 1980, when it was indeed still a crater (Ch. 5). Note the shower of molten fragments (to metres across) being sent in all directions after the bursting of a giant bubble (about 10 m wide).

Although eruptive gases are normally magmatic, explosive outbursts can be enhanced if crustal groundwater is also able to enter a conduit and, flashing into steam, can increase both the amount of magma fragmentation and the power of the escaping gases. The degree of enhancement is most pronounced when the level of magmatic explosivity is modest, as during Strombolian eruptions or the waning stages of Plinian events. Such hydromagmatic (or phreatomagmatic) activity has been commonplace in Campi Flegrei (Ch. 3), producing the distinctive cones that pepper the floor of its caldera.

Eruptive style and chemistry

Magmas evolve chemically as they cool and crystallize, the elements forming new crystals being removed from the remaining liquid. The net result is that more-evolved magmatic liquids are cooler (typical eruption temperatures declining from about 1100°C for Etnean trachybasalts to about 900°C for dacites and other evolved compositions), less fluid and, because water molecules do not readily enter most silicate minerals, contain a higher proportion of H_2O than their less-evolved predecessors. Thus, evolved magmas not only become less fluid but also have a greater potential to contain large proportions of trapped bubbles. Accordingly, among explosive eruptions the preferred eruptive style is expected to change from Strombolian to Plinian as a magma becomes more evolved (Fig. 1.4).

A simple trend between explosivity and magma chemistry is indeed easily seen among Italy's active volcanoes. Etna's trachybasalts and Stromboli's K-trachyandesites produce Strombolian explosions or lava fountains. The K-trachytes from Vulcano provide the type-example of Vulcanian activity, and the K-trachytes and K-phonolites from Somma–Vesuvius are famous for their Plinian events. At the same time, all magma types can erupt as lava flows or domes (Fig. 1.4), reflecting the general ability of migrating gases to leave large volumes of magma bubble-depleted, irrespective of its composition.

Surviving active volcanoes: a safety guide

General advice

Volcanoes are dangerous places. In the 1990s alone over a dozen volcanologists were killed "in action" and several others injured, along with dozens of journalists and tourists who ventured too close to active vents or were simply in the wrong place at the wrong time. Nevertheless, it must be stressed that, as far as the casual visitor is concerned, there is probably a much greater chance of being killed on the roads than during a visit to one of Italy's active volcanoes, provided, of course, that enough care is taken in the planning and execution of the visit. Following the deaths of six volcanologists during an explosion of the Galeras volcano (Columbia) in 1993, a series of safety guidelines for volcanologists were compiled and issued by the International Association of Volcanology and Chemistry of the Earth's Interior (IAVCEI), covering all aspects of safety on active volcanoes. Readers are encouraged to obtain a copy (*Bulletin of Volcanology* 56, 151–4; or download from http://www.iavcei.org/), particularly if they are likely to be spending any length of time on an actively erupting volcano. Six key rules are:

- Do not travel alone.
- Ensure someone locally and not on the trip knows your itinerary.
- Carry a first-aid kit, plenty of water and high-energy food for emergencies.
- Take appropriate clothing, allowing for changes in weather, and extra protection against eruptions (especially hard helmets).
- Seek expert local advice on the most recent eruptive activity.

- Be alert. A small change in eruptive behaviour can make a safe zone extremely dangerous.

Safety on the Italian volcanoes
Despite having a tendency to hurl out gas, rocks, and ash and to ooze lava, active volcanoes – particularly those of Italy – present other dangers to the visitor. They are characterized by rough and often very tricky topography, and in most cases by steep and unstable slopes. To reach the summit areas, long climbs are usually required, sometimes at considerable altitude and often in blazing sun. Appropriate footwear and clothing is therefore essential. The range of foot apparel flaunted in the summit region of Mount Etna – over 3000 m above sea level – never ceases to amaze, resulting in bleeding feet, blisters and worse. Any serious walking on the Italian volcanoes requires stout walking boots with good ankle support. The suede and Gortex type boots that now dominate much of the outdoor footwear market are ideal, combining light weight with flexibility and good resistance to damage from spiky lava flows.

Given their locations in the Naples region and points south, all the Italian volcanoes have very hot weather in summer. However, at the higher volcanoes such as Etna, Stromboli, and Vesuvius in particular, the weather can change rapidly, with thick cloud, heavy rain and thunder all possible. Consequently, although lightweight clothing is normally appropriate, a fleece or a light waterproof, or both, might also be needed. Etna, especially, presents major problems in the choice of clothing, because of its considerable altitude. Even at the height of summer, the air temperatures above 2500 m may approach freezing and, when combined with wind chill, may be significantly lower. Furthermore, the weather can deteriorate dramatically in a few minutes, with heavy snow and hail possible at any time of the year. To counter this, both warm- and cold-weather clothing should be taken, including – in the latter case – a warm fleece, cagoule or waterproof jacket, gloves, and warm headgear. Remember that Etna receives around 6 m of snow in winter and hosts the most southerly ski resort in Europe. Climbing the volcano during the winter months is not recommended unless you have mountaineering experience and are suitably equipped for alpine conditions. If the weather is fine, there is always a tendency to wear as little as possible. Remember, though, that the sun in southern Italy is extremely strong and that unprepared skin can burn and blister rapidly. Particularly at high altitudes on Etna, it is advisable to use a sun screen with a high protective factor and a hat that shades the face and neck. The scalp, nose, and tips of the ears are especially prone to burning at altitude, and the Sun may often be strong enough to burn the hands.

When the weather is hot, or if there is a strong wind (or both), dehydration can set in rapidly. Because of the permeable nature of the fractured lava flows and porous ash deposits that form them, volcanoes are notoriously dry landforms with little surface water. Enough liquid should be carried *and consumed* at regular intervals during the day.

Many of the above recommendations could apply equally well to a day's walking in any area of rugged, high altitude terrain. However, active volcanoes present other risks that even the best-prepared visitor ignores at their peril. Even when not erupting, several of the Italian volcanoes release noxious gases that can cause health problems and, albeit rarely, even death. High-temperature gases such as sulphur dioxide and hydrogen sulphide (characterized by the bad egg smell reminiscent of the school chemistry laboratory) are contained in steam emitted from fumaroles among the boiling mudpools of Solfatara in Campi Flegrei, along the rim of the active Fossa vent on Vulcano, and from the summit vents of Stromboli and Etna. The temperatures of the emitted gases range from a few tens of degrees to – in the case of Vulcano – several hundred degrees Celsius. Such temperatures can cause severe burns, and close contact with the steam should be avoided. At Solfatara and Vulcano, gas emissions are restricted to small fumarolic vents usually only a few tens of centimetres across. At Etna, there is also continuous and large-scale gas emission from the open vents at the summit. Most of the gas is water vapour at close to ambient temperatures, but it does contain sulphur dioxide and other noxious gases in small concentrations. The total amounts ejected are impressive, and even when no eruption is in progress, several thousand tonnes of sulphur dioxide per day is typically ejected by the summit vents. Sometimes the fumes from the summit vents are carried down the upper flanks of the volcano as a ground-hugging cloud. Although it may be unpleasant, inhalation of these dilute fumes is not generally a problem, although it may cause some throat irritation and could be discomforting to those who suffer from asthma and other breathing difficulties. Standing in the more concentrated fume at the summit *should be avoided at all costs* unless a gas mask with the appropriate acid-gas filters is worn.

One of the most dangerous of all volcanic gases has no odour at all and cannot be detected by the human nose. Carbon dioxide is one of the most common of all volcanic gases and it reaches the atmosphere through open vents, fumaroles and by seeping up through the ground. Because it is heavier than air, it concentrates in areas of low topography. Carbon dioxide accumulation is a particular problem around Vulcano and lethal concentrations have been detected in water wells and some buildings. Just to be on the safe side, the general rule on active volcanoes is not to rest in craters or other depressions where carbon dioxide might have accumulated.

Somma–Vesuvius has been dormant for over 55 years, and Vulcano and Campi Flegrei for over a century and four centuries respectively. In contrast, Stromboli is almost continuously erupting and Etna is rarely inactive. At both these volcanoes, extra care must be taken to minimize risk of death and injury through contact with flowing lava or ash and debris from erupting vents. Activity at Stromboli is typified by mild explosions of magma occurring every few tens of seconds to minutes, with occasional more energetic episodes when incandescent material is ejected almost continuously as a lava fountain. Although impressive, this activity is rarely dangerous or threatening *provided that it is observed from a safe distance*. Ribbons of magma ejected during the explosive blasts are torn apart to form globules of molten rock that fall to the ground as volcanic bombs. The globules cool rapidly as they fall through the air, and the plastic skin is deformed during transit to form characteristic subspherical to spindle shapes. Perfectly formed spindle bombs are something of a rarity, but don't be tempted to approach the exploding vents in search of better specimens. On reaching the summit region, watch carefully where the ejected material is falling and keep well away. Always bear in mind that the level of activity may increase without warning, ejecting incandescent debris into what might have initially appeared to be a safe zone.

Eruptive activity at Etna is considerably more varied and much less predictable than that of Stromboli, and for this reason adopting appropriate safety precautions is particularly important. As explained in Chapter 5, activity at Etna can be subdivided into two types. Major lava eruptions occur primarily on the flanks of the volcano, even at quite low altitudes, whereas both explosive activity and lava eruptions can occur in the summit region. There are currently four active vents at the summit, all capable of exploding with considerable violence and without warning. In 1979, for example, a group of volcano guides and tourists were caught by a relatively minor "throat-clearing" blast that killed nine and injured several others (Fig. 1.10). Like Stromboli, explosive activity at the summit vents of Etna may take the form of lava fountaining, although this may blast molten material to heights of several hundred metres, far higher than on Stromboli (Fig. 1.8). Large ash eruptions and sporadic explosions are also relatively common. The ash eruptions can dump tens of centimetres of ash and cinder on the upper flanks, as well as finer blankets tens of kilometres away, sometimes closing the international airport near Catania. In contrast, sporadic explosions can eject bombs and blocks metres in diameter to distances of a kilometre or more. Because of the unpredictable nature of the summit vents, they should be approached with considerable caution, even if all appears quiet. It is essential to determine the level of activity during approach and to abandon the visit if material is falling outside the vents. On occasions, one or more vents may contain small cinder cones or

Figure 1.10 A Land Rover caught by the throat-clearing explosion of Etna's Bocca Nuova on 12 September 1979. The chassis has been caved in by a block of old magma less than a metre across. Some 200 tourists were by the vent at the time: 9 were killed and more than 20 seriously injured.

be undergoing mild explosive activity. Then it is possible to get spectacular views of "red stuff" from the rims of the vents, although any apparent increase in activity should prompt immediate retreat.

When visiting the summits of Etna and Stromboli, it is strongly recommended that some form of protective headgear – such as a climbing helmet – be worn. Although this would not provide much security during a large explosion, it will offer some protection for the head against smaller or low-density fragments, or both, and hot cinder and ash. On Etna, great care should also be taken when approaching the edges of the vents, large sections of which are overhanging and unstable. On Stromboli, on no account should the vents be approached. If a blast should eject material beyond the rim of one of the vents, try not to run in panic. Look up to check if anything appears to be heading your way and take appropriate avoiding action. If all appears clear, then beat a rapid but orderly retreat.

Active lava flows on the flanks of Etna can often be observed with care from close quarters. Most flow fields on Etna consist of very spiny and rubbly aa lava that is extremely difficult to walk on and which can cause severe cuts and grazes if fallen upon. For this reason it is recommended that gloves be worn when walking over flows, even if they are cold and no longer active. Thorn-proof gardening gloves are perfectly adequate.

19

Aa flows typically crust over rapidly soon after exiting the volcano, forming levées of solidified lava along the margins of a channel that contains flowing lava. Levées are normally rubbly and highly unstable, so care should be taken when climbing them to view the incandescent and mobile channel lava. Remember that the darkened chilled crust on aa rubble may be just a thin coating around an interior still at several hundred degrees Celsius. Sometimes the main channel crusts over to form a solidified roof. Extreme care should be taken to avoid standing directly on the roof, which has a tendency to fracture and collapse back into the molten part of the flow. The temperatures of Etna's lavas reach 1100°C and can cause clothes and paper to burst into flames if they are too close. Similarly, the soles of boots may start to melt – even on solidified lava some distance from the molten part of a flow (Etnean trachybasalts are normally already solid at 950°C). If at all possible, active flows should also be observed from up wind where temperatures are considerably lower and easier to bear at close quarters. Finally, a safe distance should be maintained at all times, particularly if the lava is "spitting" as bubbles escape. Although this is not a serious problem on Etna, bursting bubbles can send small gobbets of molten rock to distances of several metres.

In conclusion, although volcanoes – especially when they are erupting – can appear to be intimidating and dangerous places, taking a few simple precautions means that the risks can be minimized and the spectacular phenomena enjoyed with peace of mind. Always:

• take care
• be observant
• take a first-aid kit.

Further reading

Websites
For general information on Italy's volcanoes and current activity, visit the sites at
http://vulcan.fis.uniroma3.it/
and
http://www.geo.mtu.edu/~boris/
Safety recommendations for fieldwork on active volcanoes can be found on:
http://www.iavcei.org/
Additional sites are given at the end of each chapter.

Literature
Cas, R. A. F., J. V. Wright 1987. *Volcanic successions*. London: Allen & Unwin.
Cortini, M., B. De Vivo (eds) 1997. *Volcanism and archaeology in the Mediterranean area.* Trivandrum: Research Signpost.
Francis, P. 1993. *Volcanoes: a planetary perspective*. Oxford: Oxford University Press.

McGuire, W. J. & C. R. J. Kilburn 1997. *Volcanoes of the world*. London: PRC.

Middlemost, E. 1997. *Magmas, rocks and planetary development*. Harlow: Longman.

Scandone, R. & L. Giacomelli 1998. *Vulcanologia*. Naples: Liguori.

Scarth, A. & J-C. Tanguy 2001. *Volcanoes of Europe*. Harpenden, England: Terra Publishing.

Turco, E. & A. Zuppetta 1998. A kinematic model for the Plio-Quaternary evolution of the Tyrrhenian–Apenninic system: implications for rifting processes and volcanism. *Journal of Volcanology and Geothermal Research* **82**, 1–18.

Wilson, L. 1980. Relationships between pressure, volatile content and ejecta velocity in three types of volcanic explosion. *Journal of Volcanology and Geothermal Research* **8**, 297–313.

Wilson, M. 1989. *Igneous petrogenesis*. London: Unwin Hyman.

Chapter 2

Somma–Vesuvius

Vesuvius is the quintessential volcano. It has produced all major styles of eruption; it has an unparalleled record of historical activity; it hosts archaeological sites of world renown; it has inspired the muse in poets and lovers; and it provides some of the best orchards and vineyards in Italy.

To travellers arriving across the Bay of Naples, the twin peaks of Vesuvius are both a lure and a menace (Fig. 2.1). Like moths to a flame, populations have been drawn to its fertile slopes and magnificent scenery. Today, the volcano supports some 600000 inhabitants and has at least another 3 million within eruptive range. Thus, while Vesuvius has been instrumental to the growth of modern volcanology, it now ranks among the most dangerous volcanoes on Earth.

Figure 2.1 The twin peaks of Somma–Vesuvius seen from Naples. The Vesuvius cone (right) has grown up within the Somma caldera, whose walls decrease in height towards the coast (right). The left peak marks the highest point on the caldera rim. Note the urban development extending from Naples to the flanks of the volcano.

The growth of Somma–Vesuvius

Vesuvius has evolved in a region forged by volcanism for at least 300 000 years. The present mountain, 1281 m high, is properly known as Somma–Vesuvius: Somma refers to the bulk of the volcanic edifice, whereas Vesuvius describes strictly the summit cone, constructed (despite episodes of collapse) during the past 1800 years (Fig. 2.2).

Consisting largely of lava flows, growth of the Somma volcano was complete before 25 000 years ago. With a summit elevation of between 1400 m and 2000 m, the volcano may well have resembled a scaled-down version of the present Mt Etna (Ch. 5). About 25 000 years ago, Somma's activity changed from modest effusions of lava to catastrophic Plinian eruptions. Five further Plinian events have occurred since then, at intervals of about 2000–8000 years (Table 2.1). The most famous is undoubtedly the most recent, which in AD 79 destroyed Pompeii and Herculaneum.

Each Plinian eruption expelled 1–3 km³ of pumice within days. The rapid rate of magma expulsion fed eruption columns that rose to 16–20 km, at which levels seasonal winds carried material inland along directions orientated from northwest to southeast. Although dustings of fine ash probably settled hundreds of kilometres away, the main blankets of ash and pumice (with thicknesses of centimetres or more) extended tens of kilometres from the volcano. Across the volcanic edifice itself, thick

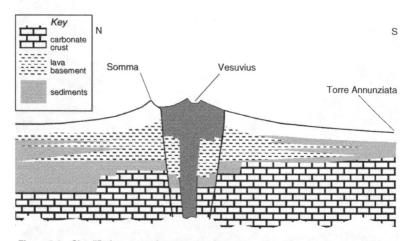

Figure 2.2 Simplified cross-section through Somma–Vesuvius. Volcanic material has punctured, and rests upon, the carbonate upper levels of the crust. Geochemical data suggest that the tops of the magma chambers feeding Plinian eruptions were about 5 km beneath the surface, just below the section. Recent seismic surveys to depths of 10 km have yet to identify fluid magma, although bodies less than 100 m across might have passed undetected. The volcanic flank deposits are 25 000 years old or younger; the lava basement is more than 25 000 years old; the sediments include volcaniclastic material.

24

Table 2.1   Plinian eruptions from Somma–Vesuvius.

Name (older names in brackets)	Date (years ago)	Deposited volume (km³)	Dense rock equivalent DRE (km³)
Codola	25000	2.2	0.7
Sarno (Basal)	17000	1.9	0.6
Novelle (Greenish)	15000	1.3	0.4
Ottaviano (Mercato)	8400	1.9	0.6
Avellino	3500	1.9	0.6
Pompeii (AD 79)	1920	2.6	0.8

pumice layers were overrun by pyroclastic flows, formed by local collapses from the unstable outer portions of eruption columns.

Rapid expulsion of magma also left unstable the walls of the emptying magma chamber, leading to inward collapse, subsidence at the surface and the formation of calderas about a kilometre across. Merged together, the collapses have created a multiple caldera 3–4 km wide. If Somma had originally reached to 2000 m above sea level, its collapse and decapitation suggest that as much as 40 km³ of the original edifice may have been lost in the past 25000 years.

The present walls of the Somma caldera are highly irregular, reaching 1000–1150 m to the north and northeast, but to only 500–700 m towards the coast. Since AD 79, the caldera floor has been buried by pyroclastic deposits and lava flows, and its centre forms the foundations of a cone with a height of about 400 m and basal width of about 2 km.

Although the early formation of Somma appears to have been dominated by outpourings of lava, effusive and weakly explosive activity was scarce between the Codola and Avellino Plinian eruptions, 25000 and 3500 years ago, and only a handful of modest eruptions (involving at most hundreds of millions of cubic metres per eruption) occurred during the following 400–500 years. In contrast, since the last Plinian outburst in AD 79, volcanic activity has been restricted to sub-Plinian eruptions and to the effusion of lava. Between 203 and 1631, at least ten eruptions (of which seven were explosive) occurred at intervals of decades or centuries, the largest being the sub-Plinian Pollena event of 472.

By 1631, a summit cone had been constructed within the Somma caldera, rising to some 50 m above the caldera rim. This construct identified Vesuvius proper, the name (apparently corrupted from the Latin *vis* for force, power or strength) denoting it as a centre of eruptive activity. Partially destroyed by the sub-Plinian 1631 event (which took about 4000 lives), the cone has since continued to grow, reaching its present height of 1281 m during the volcano's most recent outburst in 1944.

Historical activity

From 1631 to 1944, Vesuvius was in virtually continuous eruption. Activity consisted of lava flows filling up and overflowing the summit crater, punctuated by explosive summit outbursts and by occasional effusions from lateral vents in the Somma edifice (Table 2.2). Thus, although the volcano is typically associated with the AD 79 Pompeii eruption, it has been an essentially effusive centre since the seventeenth century. Indeed, the term lava was coined from the Italian *lavare* (to wash) specifically to describe the flows from Vesuvius.

Trapped to the north and northeast by the walls of the Somma caldera, lava overflows from the Vesuvius cone have been free to spill beyond the caldera only to the west, south and east (Fig. 2.3). Flank events also have been restricted to the southern and southwestern sectors of the volcano, suggesting that a major structural instability in these zones may have controlled both the location of flank fissures and the collapse of the Somma caldera. As a result, coastal towns have been most vulnerable to Vesuvius' effusions for the past three centuries.

Most outpourings have involved volumes of lava in the range 10–30 million m^3 and have continued for a matter of days. Exceptional effusions have emitted as much as 100 million m^3 over a period of years. Many rapid short-lived flows have swept the flanks of the volcano, burying settlements 5–6 km from the summit: a notable example is San Sebastiano which, in the northwest foothills of Somma, was overwhelmed three times in the interval 1855–1944.

The rapid flows have the classic aa morphology: extremely uneven

Table 2.2 Eruptive cycles, 1631–1944.

Cycle	Dates	Duration (months)
1	1631–82	c.540
2	1685–94	c. 120
3	1696–8	25
4	1700–1707	82
5	1712–37	299
6	1742–61	c.220
7	1764–7	44
8	1770–79	114
9	1783–94	135
10	1799–1822	286
11	1825–34	c.114
12	1835–9	46
13	1841–50	101
14	1854–61	84
15	1864–8	59
16	1870–72	26
17	1874–1906	387
18	1907–1944	369

(a) Somma–Vesuvius: simplified geological map

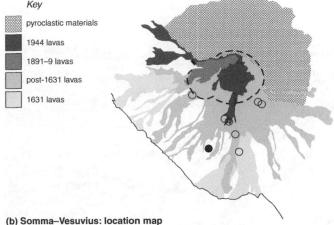

Key

- ▨ pyroclastic materials
- ■ 1944 lavas
- ▩ 1891–9 lavas
- ▨ post-1631 lavas
- ▨ 1631 lavas

(b) Somma–Vesuvius: location map

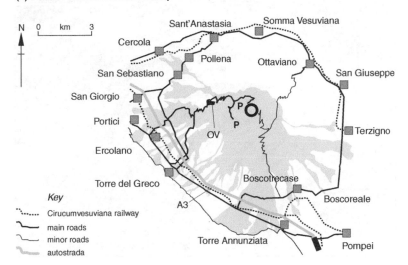

N   0   km   3

Sant'Anastasia   Somma Vesuviana
Cercola
Pollena   Ottaviano
San Sebastiano   San Giuseppe
San Giorgio   P
Portici   P   Terzigno
OV
Ercolano
Torre del Greco   Boscotrecase
Key   Boscoreale
⋯ Cirucumvesuviana railway   A3
main roads
minor roads   Torre Annunziata
autostrada   Pompei

Figure 2.3 Somma–Vesuvius. **(a)** The northeastern half of the volcano is covered by pyroclastic materials (cross-hatched), mostly from Plinian eruptions. The southwestern half has instead been almost completely resurfaced by historical lava flows, divided into the 1944 lavas (dark grey), the 1891–9 lavas that constructed Colle Umberto and Colle Margherita (mid-grey), other post-1631 lavas (grey), and the controversial 1631 lavas (pale grey). Circles show historical flank vents; the filled circle marks the pre-AD 79 Camaldoli cone. The dashed line traces the outline of the Somma caldera. **(b)** The main towns around Somma–Vesuvius and the principal connecting roads (solid lines) and the Circumvesuviana railway routes (dotted lines). The minor road (narrow line) from Boscotrecase to Ottaviano is not open to vehicles; **access by foot may also be restricted**. The Osservatorio Vesuviano museum (OV) and car-parks (stars) near the Vesuvius cone are also shown; the footpath up the cone itself starts from the northern car-park.

Figure 2.4 **(a)** The main 1944 flow shows the typical broken and jumbled surface of aa lava. The rough contorted fragments are mostly 10–30 cm across. The view is looking up the drained lava channel in the Atrio del Cavallo with the Vesuvius cone behind. **(b)** Tilted section of a ropy pahoehoe surface from the 1858 lava flow. The ropes are about 60–80 cm across and are convex in the direction of movement.

surfaces covered by irregular fragments of broken crust, normally centi-metres or decimetres across (Fig. 2.4). Very slow effusions, in contrast, have produced extensive piles of overlapping pahoehoe tongues, charac-terized by smooth surfaces and sometimes supporting ropy textures, formed by corrugations in the skin of the cooling lava (Fig. 2.4). Although individual tongues rarely extend more than tens of metres, accumulated piles have grown to about 1 km across and 100–150 m tall. Key examples are the Colle Umberto and Colle Margherita between the northwestern rim of the Somma caldera and the Vesuvius cone (Figs 2.3, 2.6). Named after the reigning king of Naples and his wife, the two lava piles were formed in succession in just less than a decade, Colle Margherita in 1891–4 and Colle Umberto in 1895–9.

The 1631–1944 activity is important not only because it is the most recent and best documented, but also because it is the only known interval of *persistent* eruptive activity since the onset of Plinian eruptions 25 000 years ago. Activity began after nearly 500 years of repose (excluding a possible phreatic outburst in 1500). Altogether, some 2–5 km^3 of magma was erupted in the course of 313 years, similar in volume to the amount of original magma (*not* erupted magma, see pp. 37–38) needed to produce a single Plinian event. Activity during this period has been grouped into 18 cycles (Table 2.2) characterized by:

- an interval of apparent repose, never longer than seven years
- a period of Strombolian and effusive activity, mostly within the crater of the Vesuvius cone, but with occasional lava overflows
- large eruptions, typically explosive (between Strombolian and sub-Plinian) or rapidly effusive (often from fissures outside the summit cone) after which a new period of repose would occur.

The notion of repose is here interesting by analogy with modern activity at Mt Etna (Ch. 5). Frequently, eye witnesses claim that a volcano is in repose provided that magma cannot be seen at the surface. Such a criterion cannot distinguish between true repose (when at least the upper parts of a volcanic system are closed) and when fresh magma is in fact residing within a central feeding conduit only a few hundred metres below ground. Thus, it is possible that magma was almost always present within the Somma–Vesuvius edifice between 1631 and 1944.

Three classic eruptions: AD 79, 1631 and 1944

The extraordinary range of behaviour at Somma–Vesuvius is exemplified by three representative eruptions: the AD 79 Plinian eruption, for which the volcano is probably most famous; the 1631 sub-Plinian eruption that ushered in the latest period of near-persistent activity, and which is the type currently believed to be the most likely should the volcano re-awaken in the near future; and the 1944 effusive-Strombolian eruption which, as well as being the most recent event from the volcano, shows the destructive potential of even the smallest of Vesuvius' outbursts.

The Pompeii eruption, AD 79

The coastline beyond Naples, from Campi Flegrei to the Sorrento Peninsula, was a prized zone for development 2000 years ago. Financed by the
Roman state, farms and villas appeared around the truncated Somma edifice, taking advantage of its fertile slopes, comfortable climate and easy
access to the sea. New economic centres were soon dominated by Pompeii,
inland to the southeast, and Herculaneum, on the coastal road from
Somma to Naples.

Although a recognized volcano, Somma had been quiet for a millennium and was thought extinct. Mild earthquakes were common, but, since
they rarely caused damage, they were not seen to be threatening. This
view changed on 5 February AD 62, when a major earthquake struck the
Somma coastline, destroying temples and villas from Pompeii to Naples.

The AD 62 earthquake was the worst in a series that were to shake the
region for the next 17 years. Although the larger earthquakes were probably tectonic in origin, some of the tremors must have been caused by
magma pushing to the surface. It is likely, too, that the volcano began to
deform, but perhaps at such a slow rate as to pass unnoticed against the
effects of the earthquakes. Certainly, no-one was prepared when Somma
burst back into life on 24 August AD 79.

The main eruption started just after midday. A column of ash climbed
more than 17 km above the summit. Blown southeast by the summer
winds, fragments of pumice began raining down on Pompeii. Within
hours, the roofs of buildings were giving way beneath piles of ash 2 m
thick. Unable to see through the ashfall, Pompeians groping their way
along the streets were caught in the deadly grip of pyroclastic flows.
Sweeping through Pompeii at 100 km an hour, the lethal mixtures of gas
and incandescent ash flattened buildings and burned lungs, erasing the
town within minutes. Forty-eight hours later, Pompeii and its people had
disappeared from sight, buried by the first ash from Vesuvius for nearly
1500 years.

Along the coastline, the sea was pulling back into the Tyrrhenian, only
to crash back onto the new beaches as a train of tsunamis. At Herculaneum, just 7 km from the summit, the terrified population was huddling
inside arches along the harbour, trapped between the volcano and the
unpredictable sea. With most of the ash raining southeast over Pompeii,
it seemed for a while that Herculaneum might escape the worst of the
eruption. It was not to be. Without warning, the town was struck by a
succession of pyroclastic flows and, within hours, Herculaneum had van-
ished beneath 20 m of pyroclastic flows and surges. The following morning, clouds from similar pyroclastic flows glided 30 km across the Bay of
Naples and into Campi Flegrei – reaching the adolescent Pliny the

Younger, who had been watching the eruption from near Capo Miseno (and from whose description such volcanic activity has become known as a Plinian eruption).

By 26 August, the eruption was spent. Once-fertile countryside had been transformed into a wasteland of smouldering ash. At least 2000 people had perished and a jewel of the Roman Empire lost for almost 2000 years. As a contemporary account describes: "Here was the seat of Venus (Pompeii); there the town of Hercules. Everything has been buried in flames, beneath squalid ashes. Not even the gods would have wreaked such havoc."

The 1631 eruption

By the mid-fifteenth century, the Vesuvian district was again a thriving commercial centre, the tree-laden hinterland around the volcano already giving way to the expansion of scattered settlements. Such success continued for the next 150 years, by which time Naples was, after Paris, the most fashionable city in Europe. Although volcanic activity had struck in Campi Flegrei in 1538 (Ch. 3), Somma–Vesuvius (now with a central cone inside the caldera) had been quiet for nearly 500 years (with the possible exception of a minor phreatic outburst in 1500).

It was therefore disconcerting when frequent earthquakes began shaking the region just before the summer of 1631. For six months the ground trembled, keeping the population awake at night along the Vesuvian coastline. During the first week of December, rumours were spreading of wells yielding discoloured water or of drying up altogether. Finally, at about 07.00 h on Tuesday 16 December, Vesuvius re-awoke with a vengeance.

Following a series of strong tremors, some felt in Naples, a fissure split the Vesuvius cone and extended WSW across the Somma caldera floor. By 08.00 h, an eruption column was rising above 13 km and sending ash in a broadly easterly direction: within 12 hours material was settling over Dubrovnik, 340 km distant; within 24 hours it had reached Istanbul, 1250 km away.

On Somma–Vesuvius itself, ash and blocks of pumice were raining down on nearby settlements. Water condensing on the ash triggered torrential rainfall, feeding floods and warm mudflows (containing newly deposited volcanic ash) that coursed down the volcano in all directions. Final annihilation arrived mid-morning on 17 December, when a succession of pyroclastic flows raced southwest to the sea (almost from the present Ercolano to Torre Annunziata), northwest beyond Cercola, and east and southeast, destroying Bosco. At the same time, ground uplift caused the sea to withdraw, sea level dropping by 3–6 m from Sorrento to the eastern edge of Campi Flegrei, later to return as tsunamis. Within 48 hours, the main explosive phase of the eruption was exhausted, only the stump of the Vesuvius cone remaining after the collapse of its uppermost 450 m. Minor activity continued to the end of the year, together with mud-flows triggered by unstable masses of ash from the volcano's upper flanks.

Altogether, the eruption killed about 4000 people and 6000 animals. Some deaths were attributable to the floods, others to the collapse of build-ings, either under the weight of accumulated ash or by the action of volcanic earthquakes. However, most casualties were caused by the pyro-clastic flows, survivors describing how they could hardly breathe as the "fiery ashes" burned the skin from their faces. The eruption also provoked the evacuation of about 40000 people, most of whom headed towards Naples. However, before the start of the eruption, warnings had been rife of another outbreak of the plague. As a result, the refugees were refused entrance to the heart of the city. Instead, processions were made showing to Somma–Vesuvius the bust of San Gennaro, the protective saint of Naples. Given the time for refugees to arrive near the city and for the bust to be carried out, the processions began shortly before the eruption began to wane, convincing the faithful of the power of their protector.

The 1631 event remains the eruption that has caused the greatest loss of life and economic damage in the Vesuvian area. Three centuries later, therefore, it is remarkable that controversy still exists as to the exact sequence of events that closed the main phase of activity on 17–18 Decem-ber. At issue is whether or not copious lava flows were effused following late-stage collapse of the Vesuvius cone (Fig. 2.3). Some eye-witness accounts suggest that, on the evening of Wednesday 17 December and throughout the following day, lava flows were effused from fissures beyond the Vesuvius cone, to sweep down towards Torre del Greco and Torre Annunziata. Such lavas can indeed be found in road cuttings near the ports of both towns. Where exposed, however, the lavas rest upon pre-1631 material, so some recent authors suggest that the outcrops in fact belong to medieval flows emplaced before 1631. The counter argument is that, at least near the coast, the flows travelled over terrain that had not been covered by pyroclastic flows, so it is inevitable that they rest on pre-1631 material (recall that, beyond the upper flanks of the volcano, most ashfall occurred to the east, away from the coast).

Since the term "lava" was coined for lava flows only during the per-sistent activity at Vesuvius *after* 1631, eye-witness accounts of the 1631 eruption use descriptive terms that appear ambiguous in the twenty-first century. In favour of the "pro-lava" camp, it must be said that some eye-witness accounts do distinguish three types of flow, which can temptingly be associated with mudflows, pyroclastic flows and lava flows. As regards lava flows in particular, the products of 17–18 December were described as bituminous and as requiring hours to reach the sea. These features are not appropriate to rapid mudflows and pyroclastic flows, neither of which is bituminous or needs more than minutes to travel the 5–6 km to the sea. Attempts to resolve this issue have been made using geomagnetic measurements to decide the age of the controversial lava flows. Unfortunately, the results have been sufficiently ambiguous for both the "pro-lava" and "anti-lava" camps to claim victory. Whatever the outcome, the controversy illustrates the scope for uncertainty in reconstructing volcanic events, even from historical eruptions that have been observed.

The 1944 eruption

By March 1944, the Italian front in the Second World War had moved north to Montecassino, some 100 km from Vesuvius on the way to Rome. Poverty and hunger were commonplace as Naples and surrounding districts began the long struggle towards peacetime stability. For more than 30 years, Vesuvius had been gently filling its crater with a succession of lava flows, sending an occasional flow down the outer flanks of the cone. Within a month of the last minor overflow in January 1944, the level of magma unexpectedly began to subside within the crater, triggering small inward collapses of its walls. If anything, the lowering of magma seemed to portend continuing tranquillity – certainly not the largest eruption for almost four decades.

Vigorous Strombolian activity began at 16.30 h on Saturday 18 March. New aa lavas filling the crater overflowed the Vesuvius cone, persistent flows heading both to the north and south. Initially advancing more than 300 m an hour, the southern flows had slowed to a halt by 21 March, their fronts nearly 3 km from the cone. However, the northern flows, blocked by the walls of the Somma caldera, headed west along the Atrio del Cavallo (the valley separating the Somma caldera from the historical Vesuvian lavas). About 2 km down slope, the flows found a notch beyond the end of the raised caldera walls and began pouring down the Somma flanks towards Massa, San Sebastiano and Cercola (Fig. 2.3). Over 12 000 people were being evacuated as the lava fronts reached the first buildings early on 21 March. By the time the fronts had halted on the following day, the outskirts of San Sebastiano and most of Massa had disappeared from sight.

Within 72 hours, the volcano had expelled some 20–25 million m³ of lava. Hopes that the eruption was waning already on the evening of 21 March were dashed when vigorous lava fountains began playing from the Vesuvius cone. (Note that the earlier lava overflows stopped being fed when fountaining began; lava advance through San Sebastiano on the night of 21 March must therefore have been the result of lava drainage from up stream.) Between 17.15 h on 21 March and 07.05 h the following morning, seven bursts of fountaining were recorded, each lasting about 30 minutes and spraying magma to heights of 2–4 km. Piles of incandescent magma accumulated around the top of the cone, eventually to flow down as small lava flows. Scoria and ash, buoyed upwards by convection above the lava fountains, were carried to the east and southeast, falling as much as 200 km away. Nearer the volcano, the blanket of ash and scoria decreased in thickness from a metre or so near the cone to about 10 cm at distances of as much as 20 km.

After a pause until noon on 22 March, the volcano began belching thick clouds of black and pink ash, rising as a pulsating brain to more than 5 km. Highlighted by lightning, the darkness of the cloud signalled the ejection of solidified fragments, indicating the start of collapse inside the crater. Weighed down by cooler fragments, parts of the billowing cloud collapsed to feed small pyroclastic flows. At the same time, giant bombs were landing at least 1 km distant, while tremors shaking the cone were triggering small landslides of the recently deposited pyroclastics. Just before 18.00 h, the volcano again paused abruptly. The calm was disconcerting, but lasted only three hours. At 21.00 h, two new eruption columns (indicating two vents) pulsed from the cone, finally terminating the paroxysmal phase of the eruption before 23 March.

The waning stage of the eruption continued for the following five days. During this time the throat of the volcano appears to have been repeatedly blocked by crater collapse, allowing accumulation of gas pressure beneath and the sudden expulsion of the blockage and underlying magma. The resulting ash clouds commonly rose to beyond 2 km. Violent shaking continued to dislodge further loose material that collapsed down the cone as hot debris flows (Figs 2.3, 2.6). The shaking had noticeably decreased by the end of 27 March and, from 29 March, activity was virtually confined to the lazy expulsion of ash following collapses within the crater.

Altogether, the eruption expelled about 35–40 million m^3 of magma (dense rock equivalent: DRE), mainly within the first 4.5 days. Despite its modest size, the outburst cost more than the buildings engulfed by the lava flows in Massa and San Sebastiano. Forty-seven people were killed, most of them crushed as roofs (as far away as Nocera and Pagano, beyond Pompeii) collapsed beneath the weight of accumulated scoria and ash. Three died at Terzigno – over 4 km from Vesuvius – having been struck by large scoria, and two were killed in a steam blast provoked by the lavas advancing through San Sebastiano. Insidiously, volcanic carbon dioxide escaping through the soil, especially from dry wells, continued to collect in pockets until the end of 1944. Many animals were killed after wandering into pockets of the gas collecting close to the ground, and parents were advised to carry babies and young children at chest height. However, a horrible example had already occurred in Ercolano on 24 March when two people in an air raid had suffocated to death in their basement shelter.

Given the massive increase in the Somma–Vesuvius population since 1944, it is easy to understand how even a relatively small eruption could provoke major damage in the future. The sequence of the 1944 event is also instructive. Conventionally, for mixed explosive-effusive eruptions, it is supposed that the lava flows appear during the final stages of activity. This sequence is most appropriate to eruptions re-opening a volcano after a period of quiescence (e.g. Vesuvius' controversial 1631 flows represented the last phase of the eruption). However, the 1944 event occurred when the upper levels of the volcano were filled with fluid magma, rather than a solid obstruction. Thus, the new gas-rich magma that fed the lava fountains had first to expel the overlying and older degassed magma that was emplaced as lava flows.

The death of a town
The slow destruction of houses in Massa and San Sebastiano was graphically described by an unnamed correspondent for *The Times* of London. The report filed on 21 March (printed on 22 March) captures the sense of helplessness when dealing with volcanic eruptions:

> The progress of destruction is almost maddeningly slow. There is nothing about it like the sudden wrath of devastation by bombing. The lava hit the first houses in San Sebastiano at about 2.30 a.m., but by dawn it still had not crossed the main street, only 200 yards away, but was nosing its way through the vines and crushing down the small outhouses more slowly than a steam-roller.
>
> As it gradually filled up the backyards of houses on the village street the flow seemed to pause. Very slowly the glowing mass piled itself up against the walls with all its weight. For a while it seemed as if it would engulf the houses as they stood but then, as the weight grew, a crack would appear in the wall. As it slowly widened first one wall would fall out and then the whole house would collapse in a cloud of rubble over which the mass would gradually creep, swallowing up the debris with it.
>
> Every now and then what looked like a geyser would suddenly spout as the molten rock engulfed a well and created a pocket of steam under pressure. Masses of steam of slightly darker quality rose as cellars full of casks of wine exploded. Over all one heard a steady cracking as the monster consumed *hors d'oeuvre* of vine stalks, olive trees and piles of faggots stored in backyards, while slowly digesting other morsels . . .

People show an apparent indifference to the disaster which is remarkable. I had expected scenes of panic, of wailing women and distracted fathers of families. There was nothing of this. Groups gathered to watch the slow immolation of the village as if it were a casual bonfire. The village doctor turned aside from saving some of his possessions to show me a good vantage point for the view.

There have even been touches of humour. We were watching the lava preparing to swallow a house which bore, somewhat unnecessarily in the circumstances, the Fascist slogan "*Viva pericolosamente*" ("live dangerously"). Presently the house collapsed. As the dust-cloud subsided a mongrel collie suddenly emerged from the masses of plaster. It shook itself and dashed to safety. It had lived up to Mussolini's injunction.

Rock chemistry and petrography

The magmas from Somma–Vesuvius belong to two broad suites: K-trachybasalt–trachyte, and K-tephrite–K-phonolitic tephrite–K-phonolite. Magmas from the second suite have been expelled since the Ottaviano eruption 8400 years ago and may also have dominated activity more than 25 000 years ago; in contrast, magmas from the Codola, Sarno and Novelle eruptions (25 000–15 000 years ago) belong to the K-trachybasalt–trachyte association.

The less-evolved magmas from both suites (K-tephrite and K-trachybasalt) are normally porphyritic, with 25–50 per cent by volume of phenocrysts. Pheonocryst contents tend to decrease as the parent magma becomes more evolved, phonolites being virtually aphyric. Major crystallizing phases are plagioclase feldspar and sanidine for the trachytic trend, and leucite, clinopyroxene and plagioclase with secondary olivine, biotite and sanidine for the phonolitic trend. The K-tephrites are associated with effusive or mildly explosive activity, whereas K-trachytes and K-phonolites tend to be found as pumice deposits from violent Plinian eruptions.

The association between chemistry, petrography and style of eruption provides clues as to the conditions of magma ascent and storage before eruption. Thus, the near-aphyric nature of K-phonolites and trachytes suggests that most crystals grew from the wall of a magma reservoir, allowing the formation of evolved magma containing few phenocrysts. As crystallization continued, volatile components (especially water) became concentrated in the residual liquid until they were able to appear as discrete bubbles. By increasing the bulk volume of magma, the growth of bubbles also increased the pressure exerted by the magma on its reservoir walls. Eventually the reservoir rock gave way to allow the eruption of a magmatic froth, the bubbles breaking the magma into a collection of fluid blobs that escaped as pumice fragments and ash.

In contrast, the strongly porphyritic nature of the K-tephrites indicates conditions suitable for crystallization throughout a magma body and not just against its reservoir walls. Their less-evolved composition, compared with phonolites, also means that they had a smaller degree of volatile concentration in the magmatic liquid upon eruption. Accordingly, the magma escaped not as a froth but as a fluid lava flow containing some bubbles, accompanied by Strombolian explosions when local conditions allowed bubbles to collect within the feeding conduit just below the surface (Ch. 1).

The phenocryst distributions within Vesuvius' lavas offer an excellent example of crystal separation under gravity. Both main phenocryst phases, leucite and clinopyroxene, can be found with centimetre dimensions. However, it is rare for such large crystals of each mineral to appear together; typically they are exclusively either of leucite or of clinopyroxene. When both phenocrysts are present in similar abundance, they are normally of modest millimetric dimensions. Phenocryst floating and sinking occurs when the density of a crystal is less than or greater than that of the surrounding magma, and it occurs at a rate that increases with the size of the crystal (hence its effect will be most pronounced when comparing large phenocrysts). The preference for large phenocrysts to consist either of leucite (less dense than the magma) or of clinopyroxene (denser than the magma) thus indicates that the parent magma ascended slowly enough for large phenocrysts to grow and to separate under gravity.

The magmatic system of Somma–Vesuvius

The chemistry of Somma–Vesuvius' products shows that the volcano has been fed by magma that can evolve from K-tephrite to K-phonolite or from a presumed K-basalt to K-trachyte before eruption. Chemical modelling suggests that the crystallization of some 5–10 km^3 of K-tephrite or K-basalt is needed to produce 1 km^3 (DRE) of K-phonolite or K-trachyte. This implies that each Plinian event must have required about 3–7 km^3 of source K-tephrite or K-basalt (Table 2.1), similar to the 2–5 km^3 of K-tephrite erupted in the period 1631–1944. Possibly, therefore, individual Plinian eruptions and the entire 1631–1944 activity were sustained by episodes of deep melting below the crust, each episode producing cubic kilometres of magma. Differences in eruptive style and frequency may thus reflect changes in the ease with which magma could reach the surface.

Separation of phenocrysts by gravity in the K-tephrites implies a slow mean rate of ascent (whether continuous or as an alternating sequence of rapid and quasi-stationary movement). The preferred range of lava volumes further suggests that magma ascends from the melting zone as discrete batches of tens of millions of cubic metres, larger effusions being fed by a collection of batches.

In the extreme, when a batch remains stalled for a very long period, it may grow into a major reservoir through the arrival of other batches from below. Crystallization along the reservoir walls eventually dominates, promoting the formation of more evolved magmatic liquid without loose phenocrysts. Compared with K-tephrite, the evolved liquid contains a smaller proportion of heavy elements (such as magnesium and iron) and so, floating upon the remaining K-tephrite, is able to collect as a lower-density aphyric liquid at the top of the reservoir. Eventually bubbles grow in the evolved liquid, raising pressures within the reservoir until the country rock fails before triggering a giant Plinian eruption.

The future

Vesuvius has been quiet for nearly 60 years. It almost certainly will erupt again: the million-dollar questions are how and when. Fortunately, these two questions may have related answers. Giant Plinian eruptions involving cubic kilometres of magma have occurred at intervals of thousands of years, whereas sub-Plinian events (e.g. 472 and 1631) involving 0.1 km^3 of magma have occurred at intervals of centuries, and the much smaller 1631–1944 eruptions (mostly of 0.001–0.01 km^3 of magma) have occurred at intervals of years. Crudely, therefore, it appears that larger eruptions can be expected after longer repose intervals, the potential volume increasing by about 0.001 km^3 for each year of tranquillity.

If the repose interval also reflects the time for which magma has been crystallizing, then longer intervals will favour the eruption of more evolved magma. Since volatiles become more concentrated within more-evolved magma, longer repose intervals also favour eruptions with a greater degree of explosivity. A simple first estimate, therefore, suggests that after a 60-year repose interval, any imminent eruption may well be explosive (or a mixed effusive–Strombolian event) and involve about 0.06 km^3 of magma. It is for this reason that, erring on the side of caution, the Italian authorities have assumed a 1631-style eruption in developing their emergency plans for the next volcanic crisis (Fig. 2.5).

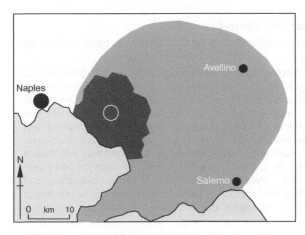

Figure 2.5
Scenario (based on the 1631 eruption) for developing emergency procedures should Vesuvius reawaken. Medium-grey zone: the amount of ashfall is expected to exceed 100 kg m^{-2}, enough to threaten roof collapse. Dark zone: vulnerable to pyroclastic flows; it covers the entire volcanic edifice, including all the towns within the excursion around Somma–Vesuvius.

Monitoring the volcano

Somma–Vesuvius is today under 24-hour surveillance by the Osservatorio Vesuviano (OV). Inaugurated on 28 September 1845, after four years' construction on the western flank of the volcano (Figs 2.3, 2.6), the OV is the oldest volcanological observatory in the world. The core of the monitoring system is a network of ten seismic stations and a Global Positioning System (GPS) geodetic array, designed to detect the increases in rock fracturing and ground uplift expected when a new batch of magma pushes its way to the surface. Information from these networks is regularly supplemented by local geodetic surveys, gravimetric studies and geochemical analyses of fumaroles within the Vesuvius cone.

Until now, nothing out of the ordinary has been detected. Even detailed seismic prospecting has failed to reveal magma at depths of 10 km or less. The conclusion is that, if the volcano is preparing for a new eruption, it is only in its early stages, when the new magma is still tens of kilometres below the surface. Observers are confident that they will know well before any eruption when Vesuvius has started again to become restless.

For logistical reasons, the operations centre of the OV is now based in the Fuorigrotta district of Naples. In April 2000, the original observatory was re-opened as a volcanological museum, its exhibits tracing the advance in volcano monitoring during the past 150 years.

Excursions to Somma–Vesuvius

Somma–Vesuvius can be visited at all times of the year. Snow (especially in January and February) can occasionally restrict access to the summit

Getting to Somma–Vesuvius

The volcano can be reached directly by road and rail, or by air and sea via Naples.

By air to Naples Naples airport (Capodichino) lies 3 km north of the city. It is served by direct services from several European cities (including London, Paris, Nice, Marseille, Helsinki, Athens and Brussels) as well from other Italian airports. Normal city buses (orange) connect the airport with the main train station, Napoli Centrale, in Piazza Garibaldi. The service runs more-or-less regularly during the day every 20–30 minutes. A more rapid coach (blue) service connects the airport with Piazza Garibaldi and, beyond that, with Piazza Municipio, in front of the city hall and a two-minute walk from the port's ferry services. Tickets for the city bus are available from the airport tabaccheria and must be purchased before boarding. Tickets for the coach can be purchased on board. Taxis are also available and, unlike the bus and coach, commonly operate through the night.

By sea to Naples Long-haul ships (traghetti) connect Naples to the Aeolian Islands and Milazzo in Sicily (departures on Tuesday and Friday; arrivals on Monday and Thursday), the Pontine Islands (daily) and Tunisia (departures on Tuesday, arrivals on Friday). Hydrofoil services (aliscafi) to the Aeolian Islands supplement the boat ferry during the summer.

By rail to Naples Naples enjoys excellent national and international railway services. Four stations may be utilized according to the route. Napoli Centrale in Piazza Garibaldi is by far the busiest station. Some through-services take advantage of the Naples metropolitan system (a mixed overground–underground link) to cross the city, stopping at one or more of the following stations: Napoli Campi Flegrei (in the city's western suburbs), Napoli Mergellina (the west-central part of the city) and Napoli Piazza Garibaldi (directly beneath Napoli Centrale in the east-central part of the city).

By rail to Somma–Vesuvius Somma–Vesuvius is serviced by the Circumvesuviana, a local railway link with lines running along the coast from Naples to Sorrento and inland from Naples to Sarno. The coastal line has stations at Ercolano and Pompeii for visits to the adjacent archaeological sites. A bus to the Vesuvius cone runs daily from the forecourt of Ercolano station. Some long-haul trains stop at Torre Annunziata, on the southwestern fringe of the volcano.

By road to Somma–Vesuvius The southwestern (coastal) flank of the volcano is most easily reached by the A3 motorway. Exits at Ercolano and Torre del Greco are best for visits to the summit. (Torre del Greco is more suitable if travelling by coach). The northern and eastern (inland) flanks are best reached via the SS268 or the motorways A16 (north flank) and A30 (eastern flank).

cone. During the peak season (July and August), it is common to be sandwiched between coaches on the summit road.

Somma–Vesuvius became a national park in June 1995, a major step in the bid to halt illegal construction and general environmental degradation of the volcano. Since then, whole tracts of the mountain have been cleared of waste, and new paths have been established for visitors wishing to enjoy its local flora and fauna, as well as its geology. The remit of the national park is to conserve the unique ecological status of the volcano. Previous conservation measures had been restricted to sectors in the control of the Corpo Forestale (the Italian forestry commission). By necessity, many of these sectors had to be fenced off from the public. In 2000, the Corpo Forestale formally handed control of its sectors to the national park, so that greater public access can be expected from 2001. However, until the changes have been established routes have been chosen here that *ought* not to require prior permission for access. **Note that this does not constitute a guarantee of free access**. For the latest information on access, contact the headquarters of the national park at San Sebastiano on 081-771.7549.

From the sea to the summit

The western ascent of Somma–Vesuvius follows the only route to the summit available to vehicles, including public transport. It illustrates the striking contrast between the overdeveloped lower slopes and the wilder upper flanks (excluding the tourist car-park near the summit), and is the most evocative route for understanding the impact of eruptions on human activity since 1631.

Leave the A30 motorway at the exits for either Ercolano or Torre del Greco and follow the signs to Vesuvius (the same roads can be reached on foot from the corresponding Circumvesuviana stations). From either exit, the first 3–4 km (to about 350 m above sea level) is disappointing geologically. The roads cross an ugly hotch-potch of urban development and decrepit orchards and vineyards, most of which are excellent examples of

Excursion time Three to four hours by car; about six hours by public transport; and 10–12 hours by foot. Allow 1–2 hours at the summit cone itself.

Public transport A bus service runs from the Circumvesuviana station at Ercolano to the summit, stopping for about 45–60 minutes on the way up at the base station of the now-inactive chairlift (seggiovia) at the bottom of the Vesuvius cone. The timetable varies according to the season. Check tourist information offices for details or ring the Circumvesuviana on 081-772.2444.

Maps Istituto Geografico Militare topographic maps: 1:25 000, Sheets 448/III/ Ercolano, 466/IV/Torre del Greco. Touring Club Italiano 1:50 000 topographic map, Il Golfo di Napoli, 1. Consiglio Nazionale delle Ricerche, 1:25 000 geological map of Somma–Vesuvius. These maps are often difficult to find, even in Naples.

Difficulty Moderate for the Vesuvius cone, which can be climbed slowly in 20–30 minutes. A path has been made from compacted 1944 scoria; stout shoes are advisable. For those walking from the coast, the route to and from the Vesuvius cone is easy but very long, so water and food must be taken.

how not to construct discreetly in a district once perceived to be of exceptional natural beauty.

The routes from Ercolano and Torre del Greco eventually join at a T-junction, recognizable for its cluster of restaurants and, on busy days, for the habit of patrons to use the central part of the T as an unofficial car-park. Those arriving from Ercolano continue ahead; those from Torre del Greco turn right. Above this junction (at 350 m above sea level) the visitor appears to be on another volcano. Suddenly the winding road is flanked by outcrops of dark grey lava, appearing haphazardly through the grass and broom (Fig. 2.6). For the first 1.5 km, the road twists mainly across the 1858 lava flows, notable for their tracts of ropy pahoehoe, before reaching

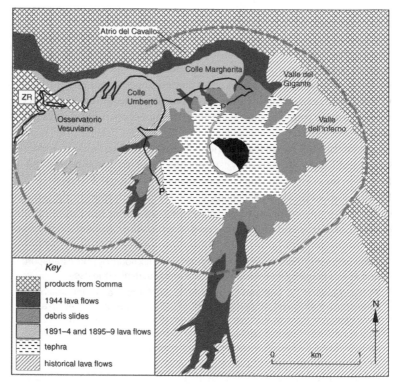

Figure 2.6 Principal features in the summit region of Somma–Vesuvius. The region is bordered to the north (top) and east by products from Somma (cross-hatched) and, elsewhere, by historical lava flows (white). The road to the summit (solid line) passes the bar-restaurant Zi Rosa (ZR) and the Osservatorio Vesuviano museum (OV), as well as Colle Umberto (CU) and Colle Margherita (CM), the respective sources of the 1895–9 and 1891–4 lava flows (pale grey). The 1944 lava flows (dark grey) are overlain on the Vesuvius cone by tephra (dashed) and debris slides (medium grey). Within the summit crater, the southwest dotted area marks the exposed section of the pre-1906 crater; the vertical lines denote exposed products from the interval 1913–44. The stars locate summit car-parks and the ascent up the cone is shown by the small dashed line. The large dashes outline the trace of the Somma caldera.

the bar-restaurant Zi Rosa on the left side of a sharp right-hand bend at 550 m above sea level. As well as offering the least-extortionate refreshments on the upper flanks of the volcano, Zi Rosa overlooks a superb view of the main 1944 lava flow – which appears as a mass of grey rubble covered by the greenish lichen *Stereocaulon vesuvianum* – that down slope destroyed Massa and parts of San Sebastiano.

Virtually opposite Zi Rosa is the side road leading to the original Osservatorio Vesuviano. In view of restricted parking space close to the OV, it may be easier to leave any vehicle by Zi Rosa and walk the 500 m to the observatory (take the left fork at the junction after 300 m).

The OV museum is open to the public at weekends (*c.* 10.00–13.30 h) and by appointment for groups during the week (081-610.8483). Looking east, the OV offers wonderful views of the Vesuvius summit cone and the reforested flanks beyond. The sharp line up the right half of the cone marks the path of the chairlift inactive since the 1970s. Built on a raised promontory of pyroclastic material, the OV stands on an island flanked either side by the now-vegetated lavas issued during the growth of Colle Umberto in 1895–9. Away from the volcano, the OV affords good views to Capri and the Sorrento Peninsula.

From Zi Rosa the main road continues to cross the 1895–99 flows, with the 1944 flow close on the left. Both sides of the road have been fenced in by the Corpo Forestale, denying access particularly to the 1944 flow; this situation may change in 2001, so look out for recent official entrances on the left-hand side during ascent. After about 2 km of sharp bends, the road curves to the right around the tree-covered Colle Umberto (on the right) before reaching a major fork. To the right, the road continues for 2 km, crossing 1944 lava flows and debris slides and ending at the base station (together with forlorn bar) of the defunct chairlift up the Vesuvius cone. The car-park provides a good panoramic view of the coast.

The left fork leads to the Vesuvius cone, crossing a steep incline covered by the 1891–4 lava flows from Colle Margherita. On the right (towards the Vesuvius cone) the road passes the snouts of what appear to be three steep-sided aa lavas, metres thick and tens of metres across. These are in fact the landslides (or debris slides) triggered by shaking during the final stages of the 1944 eruption and they consist of remobilized aa and scoria fragments. The surface material is lava grey and covered by lichen, whereas internal fragments are frequently ochre red.

To the left of the road, the slope dips steeply towards the main 1944 lava flow within the Atrio del Cavallo ("horses' courtyard"), the area where visitors once left their horses and donkeys before ascending Vesuvius. The Atrio marks the western entrance to the depression between Vesuvius and the upstanding walls of the Somma caldera. Moving east, the Atrio opens

Figure 2.7 The inner walls of the Somma caldera seen from the road approaching the summit car-park of the Vesuvius cone. The resistant pale-coloured dykes form ridges jutting from the main wall.

into the Valle del Gigante ("valley of the giant") which, curving southeast, merges with the Valle dell'Inferno ("valley of hell"). The sheer walls of the Somma caldera offer stupendous sections through the inside of a volcano (and strongly resemble views along the walls of Etna's Valle del Bove; Ch. 5). They reveal as much as 300 m of accumulated pyroclastic deposits, erupted by Somma's Plinian events. At 1134 m above sea level, the highest point visible is also the highest part of the Somma caldera; known as the Punta del Nasone ("tip of the big nose") it is easily identifiable.

Lava dykes crosscut the pyroclastics at mostly subvertical angles, some jutting out in relief towards the observer (Fig. 2.7). The dykes show a radial arrangement about a centre that must have been located somewhere near the present Vesuvius cone. On occasion, a subvertical dyke can be seen to become narrower and to pinch out towards both top and base, suggesting that it was propagated laterally from a central axis and not vertically from below: in other words, it was possible for fluid magma to have been injected laterally into the flanks of Somma from a central feeding conduit. This indicates that lava could reside within the upper levels of the volcano, suggesting that effusions may not necessarily have been unusual between Somma's Plinian eruptions.

Some 150 m below the Vesuvius car-park, the road bends sharply to the right next to the ruins (on the left) of an OV geophysical station. It is common practice to park along this final stretch of road, in order to avoid the fee of the official car-park; however, vehicles are left there at the user's risk.

From the car-park, it is possible to follow the upper reaches of the main 1944 flow along the edge of the Somma caldera walls and, although it has been obscured by later scoria, the trace of the overflow that fed the lava can be imagined passing up the flank of the Vesuvius cone just outside the car-park entrance. Notice how drainage has depressed the central part of the main flow and left raised levels both against the caldera wall and also at the upstream end of the flow, where it bends sharply to the west (left) in the Atrio del Cavallo. Drainage occurred after the flow had stopped being fed from the summit and explains why lava fronts continued to advance through San Sebastiano even though summit activity had entered a phase of violent lava fountaining.

The track up the Vesuvius cone begins at the far end of the car-park. The cone is covered by scoria and lithic fragments ejected during the waning stages of the 1944 eruption. Looking down slope, the lobate edges of the 1944 lava apron can be distinguished at various locations beyond the base of the cone. The late-stage debris slides are also visible as raised tongues. Following the path of the slides up the cone, it is possible to identify on both sides of the path some of the arcuate scars left behind at the head of a collapse. Beyond the cone itself, the ascent offers excellent views of the OV, Colle Umberto and the coastal flank of Somma–Vesuvius.

Arrival at the summit is marked by a ticket office. Some pressure may be given (especially to groups) to be accompanied by a guide. In practice, you can accept or decline the offer at your discretion, but a gratuity is expected if you agree to be accompanied (notionally 10000–50000 lire, increasing with the size of groups up to 10–20 people).

The dramatic first sight of the cone interior, immediately to the left beyond the ticket office, reveals a gaping hole almost 500 m across at its widest and 320 m deep (Fig. 2.8). The far wall consists of innumerable lava layers, each typically metres thick, capped by a thick pyroclastic layer. The lavas mark stages in the gradual infilling of the crater between 1913 and 1944. They demonstrate that the crater was filled not by a growing lava lake (with a continuous, molten interior) but by the repeated overflow of lava streams from a narrower central feeding system. The base of the crater moved up as each flow spread and solidified over its predecessor. As described by eye witnesses, small spatter cones commonly grew and collapsed over the feeding vent.

The main 1944 lava overflowed the crater somewhere behind the ticket office. The thickness of the lava itself is difficult to determine from a distance, owing to the superposition of giant blebs of magma emplaced by the lava fountains on 21 March 1944. Still fluid upon landing, some blebs accumulated and began to flow as a lava, thereby blurring their contact with the lava overflows active from 18 to 21 March. Notice also that, with

Figure 2.8 The northeastern wall of the Vesuvius crater. **(Above)** Close-up (taken from observation point by ticket office) of the lava layering produced by the 1913–44 infilling of the crater. Individual lavas are metres thick. The discontinuity dipping left to right on the lower left marks the junction between the pre-1906 (left) and pre-1944 (right) crater walls. **(Opposite)** Subhorizontal layering above the scree marks the succession of lava flows that filled the crater during the interval 1913–44. The section is covered by 1944 pyroclastic material, the low resistance of which to erosion is shown by the funnel-shape structures.

the accumulation of fountain ejecta, the present northern rim of the cone is no longer the low point that it had been before the 1944 event.

The fine nature of the pyroclastic material covering the far side of the cone is demonstrated by how easily the unit has been eroded to yield prominent downward-tapering funnels (Fig. 2.8). Most of this unit was emplaced during the paroxysmal phase of the eruption on 22 March.

An irregular subvertical discontinuity can be seen in the northern wall of the crater, just to the left of the observation point by the ticket office (Fig. 2.8). A similar discontinuity runs down the opposite southeastern wall. The discontinuities are a result of how the interior of the crater collapsed at the end of the eruption. To the north and east, the collapse left behind remnants of the 1913–44 lava infilling, whereas to the south and west it cut farther back and into the wall of the crater left after the 1906 eruption, whose final explosive phase was similar to that of 1944. The discontinuities mark the contacts between the craters formed by collapse at the ends of the 1906 and 1944 eruptions.

The summit path continues to the southeastern rim of the crater. A path once continued beyond the crater, but has been left unmaintained for several years. The guides may prevent access beyond the summit. However, it is still worth walking as far along the rim as possible, not only for a better feel for the size of the crater, but also for a good overview of the eastern flanks of the volcano. With binoculars, it is possible to discern the ruins of Roman Pompeii, just inland from Torre Annunziata to the southeast (to start with, it is easiest to look for the distinctive shape of the amphitheatre).

While on the southern rim, collect a handful of surface material from the outer side of the cone. Apart from irregular scoriaceous fragments millimetres or centimetres across, the surface contains abundant euhedral pyroxene crystals which, dark green to black and millimetres long, have been completely separated from any enclosing fluid magma. These represent some of the last material to have been ejected by the 1944 eruption and may correspond to crystals stripped from the walls of the feeding system. In contrast, leucite phenocysts are very scarce. The relative proportion of pyroxene and leucite crystals is notable when comparing their abundances in the massive lava (whether from lava overflows or lava fountains) found in outcrop around the crater rim. These lavas (as well as the main 1944 flow in the Atrio del Cavallo) typically contain as phenocrysts 15–20 volume per cent leucite, 10–15 per cent clinopyroxene (some with paler green olivine cores) and about 10 per cent plagioclase feldspar.

Chemical analyses of the 1944 products show a progressive increase in magnesium and decrease in potassium during the course of the eruption. Most of the changes occur in products erupted after the initial lava effusion and can be related to an increase with time in the relative proportion of magnesium-rich pyroxene compared to potassium-rich leucite. Before reaching the surface, magma feeding the eruption must have resided within the crust for a period long enough to allow some separation of leucite and pyroxene phenocrysts under gravity.

Into the Valle del Gigante (medium difficulty, boots advisable, water essential; time, 1–2h)

Although the Vesuvius cone is fenced off along the final approach to the summit car-park, the fencing moves away from the road at the sharp bend by the remains of the OV geophysical station. Here the car safety barrier can be crossed and a path found leading from the ruins down to where the 1944 flow reaches the Atrio del Cavallo. It is then straightforward to cut across to the Somma caldera. **Beware of rockfalls when close to the caldera walls.** Return to the road by the same route, looking out for olivine–pyroxene nodules thrown out by the 1944 eruption.

When the national park takes active control of the summit region, it will be possible to have access from here, through the Valle del Inferno and back to the starting point along the inland flank of the Vesuvius cone (a good day trip). For the latest information, the headquarters of the national park at San Sebastiano can be contacted on 081-771.7549.

Around the volcano

Local roads join to form a circular route 25–30km around the base of Somma–Vesuvius. The route offers the opportunity of exploring the bulk geomorphology of the volcano, examining stratigraphic sections through Somma's Plinian deposits, visiting archaeological sites, and contrasting the human development of Somma on its coastal and inland flanks.

Excursion time. The whole route can be completed by private vehicle within a day, excluding the time needed to visit the Roman sites (see below) of Herculaneum (Ercolano), Pompeii (Pompei) and Oplontis (Oplonti). Alternatively, selected stops can be chosen for half-day visits by public transport. Some sites can be reached by foot from Circumvesuviana railway stations; all can be reached using rural buses (check timetables on arrival).

Maps. As for the western ascent of the volcano.

Difficulty. Easy: most walking is along roads. There are no steep inclines near the main stops.

Since the route is circular, it can be started where convenient. The description here begins at the A30 motorway exit at Torre del Greco and follows a clockwise direction around the volcano (Fig. 2.3).

From the Torre del Greco exit, turn right up hill and then left at the first crossroads (after about 250 m). Follow the road for about 5 km until reaching San Sebastiano. On the way, look up hill to the right for panoramic views of the Vesuvius cone, Colle Umberto and the Somma caldera rim, which from some directions appear as three peaks. The road rises over the 1944 lava flows when passing through San Sebastiano and Massa. Although residential development has expanded both towns in the past 30 years, outcrops of the 1944 lava can still be seen where incorporated into the foundations of buildings.

Continue about 1 km beyond Massa to Pollena–Trocchia. From here, the volcano appears as a single flat-top structure, the Vesuvius cone being hidden by the Somma caldera. The dominance of pyroclastic material within the edifice is demonstrated by the deep radial incisions cut by water into the mountainside. From the corner opposite the church in Pollena's main square, a small road runs up towards Somma. After about 1–1.5 km, the road passes a disused quarry on the right. This area is currently being reclaimed by local environmental groups, so vehicle access into the quarry may be restricted. Pollena can be reached by bus or from the Trocchia Circumvesuviana station, less than 1 km north (away from the volcano) from Pollena's main square.

Quarrying has revealed an almost complete stratigraphic sequence of Plinian deposits from Somma since the Sarno eruption 17000 years ago. An idealized section is given in Figure 2.9. A useful marker horizon is the sequence of white–black–white pumice about half way up the sequence. Originally attributed to a Somma eruption (called Lagno Amendolare), the pumice has since been recognized as fallout from an eruption in Campi Flegrei about 11400 years ago.

Access to the quarry is restricted by vegetation and topography. Pyroclastic flow units can be recognized by their basal layers rich in angular lithic fragments (mostly black lava, with some white pieces from the

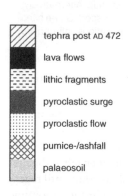

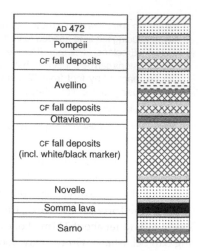

Figure 2.9 Simplified stratigraphy at the Pollena quarry. Deposits from the Sarno to Pompeii Plinian eruptions from Somma are interspersed with palaeosoils and fall-deposits from eruptions in Campi Flegrei. Note the lava (and Strombolian cone) present between the Sarno and Novelle eruptions. The thickness of the whole sequence varies from about 70 m to 90 m.

underlying carbonate basement; Fig. 2.2); some flows have also eroded and infilled channels cut into the pyroclastic levels beneath. Pyroclastic surges can be recognized (with practice) by faint layering at the centimetre scale; occasionally the layering is cross bedded. Pyroclastic-fall units maintain similar thicknesses over irregularities in the underlying terrain; many are graded, material normally becoming finer towards the top, but this may not be obvious until close to the outcrop.

About 100 m up hill from the first entrance to the quarry, excavation has revealed a Strombolian cone with feeding dyke and lava flows, emplaced some time between the Sarno and Novelle eruptions (Table 2.1). Red horizons immediately beneath the flows show where the heat of the lava has baked and oxidized the Strombolian scoria. The outcrop confirms that effusive activity occurred between the earlier of Somma's Plinian eruptions, and suggests that some of the dykes that propagated outwards from Somma's axial conduit may have fed low-altitude flank eruptions (see excursion to the summit). It is notable also because it records an outburst from the volcano's northwestern flanks, precisely the sector untroubled by flank activity since at least the AD 79 eruption.

Returning to the main square, turn right (east) and follow the road around Somma through Sant'Anastasia, Somma Vesuviana and Ottaviano. Time permitting, it is worth stopping at Somma Vesuviana to explore the sixteenth-century core of the town and to compare this with more recent urban development. Ottaviano is remarkable from a volcanological viewpoint because, during the 1906 eruption, bombs from Vesuvius systematically shattered windows on the side facing *away* from the volcano. This curious effect was the result of uphill winds, generated by the heat of the eruption column, that were strong enough to suck projectiles back towards the volcano.

From Terzigno, 3 km south of Ottaviano, the view west reveals the true extent of the Somma caldera, which appears as a flat top about 4 km wide with the Vesuvius cone emerging from just right of centre (Fig. 2.10). About 500 m south of Terzigno, the road passes for nearly 1 km over the 1834 lava flow. Immediately beyond the flow, two roads merge from the right. Turn right and follow the right-hand road for about 500 m, until reaching the entrance of the Cava Ranieri (Ranieri Quarry; note that this is not the same as the "C. Ranieri" marked nearly 2 km to the south on the 1:25 000 maps). You will know if you have missed the turning if you arrive at the hamlet of Caprai al Mauro, 500 m farther south. Those travelling by public transport should ask for either Caprai al Mauro or Boccia al Mauro (just 300 m to the east) and walk the kilometre or so northwest to the quarry.

Cava Ranieri is a working quarry (Monday–Friday and Saturday morning) and permission for access should be obtained from the site manager. The quarry is one of the few with a permit to excavate the massive interiors of lava flows (in this case the 1834 flow) for use as building material. Roman ruins excavated from the AD 79 deposits are on the right when following the main track through the quarry to the site offices. Unless blasting is taking place, the company is normally relaxed about small groups of ten or fewer visiting the site. You can try your luck on the day, or phone in advance (especially for large groups) on 081-529.6552.

Before visiting the Roman site, it is worth looking at the exposed lava (Fig. 2.10). The aa flow has a classic brecciated top and base, separated by a massive interior up to several metres thick. Stretching as far as the eye can see, it is easy to imagine the fear provoked by such a flow rolling forwards while still incandescent. At the main quarry face, it may still be possible to see what appears to be a dyke, about 1 m thick, feeding the flow. On either side of the dyke are pieces of surprisingly well organized breccia. However, this is not an exceptional example of magma blasting its way to the surface, but of lava filling an ancient well, lined with bricks.

The Roman remains themselves consist of a small country villa. Excavation has revealed a 2 m-thick sequence from the AD 79 eruption (Fig. 2.11). Although in a quarry, this is an archaeological site and **no samples must be taken**. The section illustrates the evolution of virtually the entire eruption (see the Oplonti section below for a discussion of magma chemistry and petrography):

- A basal metre of white pumice and ash (typically 1–10 mm across), corresponding to fallout from the eruption column on 24 August AD 79. The unit rests on a palaeosoil, indicating the long period of repose since the previous major eruption from the volcano.
- About 70 cm of pyroclastic flow and surge units, containing material of typically submillimetre dimensions. These were produced by a succession of collapses from the outer, denser part of the eruption column.

Figure 2.10 **(a)** Profile of Somma–Vesuvius looking west from the Ranieri quarry. The snow-capped cone of Vesuvius rises from the Somma caldera, whose flat rim can be seen sloping gently downwards to the right. The 1834 lava caps the quarry face in the foreground. **(b)** The 1834 lava flow, looking almost up stream. The massive lava interior is quarried for building stone.

They probably indicate a decline in the rate of eruption, since a reduced eruption rate would have decreased the rate of atmospheric heating by the magmatic fragments, thereby reducing the buoyancy carrying the fragments upwards and favouring column collapse. A declining eruption rate would also reduce the distance to which large fragments could be carried from the volcano (whether in a fall or flow unit) and may thus be the factor underlying the decrease in particle size compared with the earlier ashfall units.

- A layer about 15 cm thick containing angular lithic fragments, presumably pieces from the walls of the feeding system. The eruption is waning and the magmatic system is collapsing in on itself.
- About 10 cm of accretionary lapilli, characterized by subspherical pisolites millimetres across. A pisolite has a concentric layered structure and grows as several coats of wet fine ash cover a tiny fragment. It can form only when the eruption column contains large amounts of moisture. The required moisture may have come from meteoric water filtering into the magmatic system when the conduit walls began to collapse.
- A top layer of about 30 cm of pyroclastic surges. These represent the final stages of eruption on 25–26 August. The eruption rate was too slow to sustain an eruption column, which episodically collapsed to produce the surges. Among these surges, in particular, it may be possible to distinguish stratification and cross-bedding features.

Returning to the main road, turn right and continue through Passanti (2 km) towards Boscoreale. The road skirts just south of Boscoreale and after another 3–4 km reaches the coast at Torre Annunziata, host to the villa di Poppea, the principal building excavated from the village of Oplonti, also destroyed by the AD 79 eruption. At the northeastern edge of Torre Annunziata (about 500 m southeast from the A30 exit and about 1 km southeast from the Circumvesuviana station), the entrance to the villa opens onto a side road running perpendicular to the coast. Parking nearby is restricted and it may be necessary to leave vehicles in adjacent streets. **Remember that this is an archaeological site and sampling is not permitted**.

The villa di Poppea is among the best preserved of all the Roman buildings overwhelmed by the AD 79 eruption and 1–2 hours should be allowed for a visit to the archaeological site alone (tel.: 081-862.1755). The key stratigraphic sections can be seen immediately in front of the villa. The amount of exposure varies according to the degree to which plants have been allowed to run wild. The essential upward sequence (Fig. 2.11) is similar to that at the Ranieri quarry: pumice fall, ashfall, pyroclastic flows, lithic fragments, and pyroclastic surges and accretionary lapilli. However, at about 5 m the total thickness of the sequence is more than twice that of the

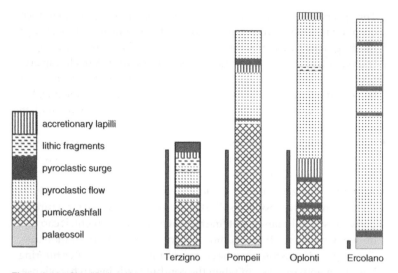

accretionary lapilli

lithic fragments

pyroclastic surge

pyroclastic flow

pumice/ashfall

palaeosoil

Terzigno   Pompeii   Oplonti   Ercolano

Figure 2.11 Simplified stratigraphic sections through the AD 79 deposits. All the sections are found next to archaeological excavations, so access may be restricted. The sections Pompeii and Ercolano are from the main tourist sites. Terzigno is from the country villa just beyond the main entrance to the Ranieri quarry (permission for access required from the quarry manager). Oplonti (villa di Poppea) is in Torre Annunziata. Note that pumice and ashfall deposits are found only at sites to the south and east of the volcano; the Ercolano deposits consist of pyroclastic flows and surges. The vertical bars represent thicknesses of 2 m.

Ranieri Quarry outcrop. Since both locations lie at similar distances from the volcano, the greater thickness of the Oplonti sequence demonstrates that the villa di Poppea had been closer to the axis of the eruption cloud as it was blown southeast (Fig. 2.12).

In contrast to the white pumice at the Ranieri quarry, the basal pumice at Oplonti changes upwards from white to grey. The difference in colour reflects a change in both chemical composition and petrography. Whereas the white pumice is an aphyric K-phonolite, the grey pumice is a porphyritic K-tephritic phonolite (less evolved and richer in magnesium; Ch. 1) with millimetre-size phenocrysts mostly of clinopyroxene, biotite and alkali feldspar. The white pumice is less dense than the grey component, consistent with a chemically stratified magma chamber from which the less dense upper levels of K-phonolite were erupted first.

The villa itself reveals the dramatic impact of pyroclastic flows on structures. The northern corner of the building (the first corner reached after descending the ramp from the ticket office) is lined by a series of broken pillars. Most have been broken at similar heights above their bases. The height of fracturing corresponds stratigraphically with the change from fall to flow products. Partially buried by the initial fall of pumice and ash, the exposed upper parts of the columns were struck by a series of

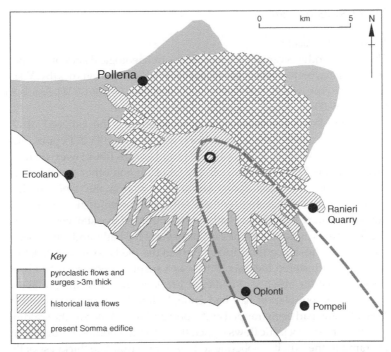

Figure 2.12 Distribution of AD 79 products. Seasonal winds blew the ash cloud southeast-wards over Pompeii. The dashed line encloses the area with ash deposits thicker than 1 m.

pyroclastic flows racing at hurricane velocities. The blast either from the flows or from the airwaves that the flows pushed in front of themselves was sufficient to snap the columns in two. Now imagine what would happen to a human being.

The Torre Annunziata entrance to the A30 is easily reached from the villa. (Before returning, those curious about the 1631 eruption may like to go the port and look at the lava flows exposed along the back of the so-called lido azzurro. Are they from 1631 or not?) Taking the Naples direction, the motorway cuts through several aa lava flows and passes three flank cones on the right. The first is medieval, the second was formed in the 1760 eruption, and the third (most westerly) is Camaldoli, a pre-AD 79 vent now host to a monastery. The exit to Torre del Greco is reached after about 8 km and circumnavigation of the volcano complete.

Other sites to visit

Pompeii and Ercolano

The AD 79 products can also be seen at the archaeological sites of Pompeii and Ercolano, both of which have Circumvesuviana stations nearby. Visits to each site may take at least half a day (Pompeii tel: 081-861.0744; Ercolano tel: 081-759.0963). Again, **sampling is not permitted**.

Pompeii being southeast of Somma–Vesuvius, the exposed AD 79 sequence resembles those at the Ranieri quarry and villa di Poppea (Figs 2.11, 2.12). In contrast, pyroclastic flows and surges (ibid.) overwhelmed Ercolano. The differences between the Ercolano and other three sections emphasize the difficulties in correlating deposits from the same eruption.

At Ercolano, archaeologists have left intact some of the pyroclastic flows that entered the Roman baths at the southeast (coastal) corner of the site. The pre-eruption coastline ran just beyond the baths and, underneath a series of arches along the adjacent port, in the 1980s dozens of skeletons were found clustered together. This was a surprise because, it had been assumed that the absence of bodies at Ercolano indicated that, unlike at Pompeii, the population had had enough time to escape the eruption. Instead many had been trapped at the port and were killed as the scalding heat from the pyroclastic flows ripped through their bodies. Returning up the ramp to the exit, it is sobering to look at the complete thickness of the flow deposits and to realize that they were emplaced at most within hours.

Further reading

Websites

For the Vesuvius National Park:
   http://www.cts.it/parchionline/pnvesuv/arrivare.htm
For the Vesuvius Observatory:
   http://www.osve.unina.it
For the history of Somma–Vesuvius and other Italian volcanoes:
   http://vulcan.fis.uniroma3.it

Literature

De Vivo, B., R. Scandone, R. Trigila (eds) 1993. *Mount Vesuvius*. Special issue of *Journal of Volcanology and Geothermal Research* 58 [whole volume].

Nazzaro, A. 1997. *Il Vesuvio*. Naples: Liguori.

Santacroce, R. (ed.) 1987. *Somma–Vesuvius*. Volume 114, Quaderni de "La Ricerca Scientifica", Consiglio Nazionale delle Ricerche, Rome.

Sigurdsson, H. S. Carey, W. Cornell, T. Pescatore 1982. The eruption of Vesuvius in AD 79. *National Geographic Research* 1, 332–87.

Spera, F. J., B. De Vivo, R. A. Ayuso, H. E. Belkin (eds) 1998. *Vesuvius*. Special issue of *Journal of Volcanology and Geothermal Research* 82 [whole volume].

Chapter 3

Campi Flegrei (Phlegraean Fields)

Campi Flegrei is the fiery birthplace of myth and legend. It was here that Hercules, Ulysses and Aeneas faced some of their greatest challenges; that the Cuman Sibyl released her ambiguous prophesies; that the Ancients placed the entrance to Hades; and that the Earth was shaken by one of the largest eruptions in the Mediterranean.

Evolution of Campi Flegrei

Tucked between Naples and the Tyrrhenian Sea, Campi Flegrei (from the Greek "burning fields") is a giant caldera 12–15 km across that has been the site of activity for at least 50 000 years. Conventional wisdom suggests that the main caldera formed about 35 000 years ago when, to the terror of Palaeolithic settlers, one or more eruptions spewed out the Campanian Ignimbrite (Ignimbrite Campana), at least 100 km^3 of pumice that covers more than 30 000 km^2 (controversial aspects of this interpretation are discussed below). This caldera was in turn shattered some 23 000 years later by another caldera-forming event, this time emplacing the relatively modest 20–50 km^3 of the Neapolitan Yellow Tuff (NYT; Tufo Giallo Napoletano). About 6 km across, most of the NYT caldera today lies under water in the Bay of Pozzuoli.

For the past 12 000 years, activity in Campi Flegrei has occurred in a ring 3–4 km wide around the rim of the NYT caldera (Fig. 3.1). With typical volumes of 0.1–1 km^3 (Table 3.1), the eruptions have been separated by intervals of the order of centuries or less, apart from hiatuses (at least across the present subaerial part of Campi Flegrei) between about 8000 and 4500 years ago and between 3000 years ago and AD 1538. The first interval was marked by the erratic uplift of the Starza marine terrace, the edge of a block 5 km long that runs along the coast from Solfatara to Monte Nuovo (Fig. 3.1). The terrace today stands about 40 m above sea level.

Table 3.1 Principal eruptions in the formation of Campi Flegrei.

Name	Date (years ago)	Deposited volume (km³)	Chemistry
Lava domes; eruptions from Pròcida	48 000–37 000		Domes: K-phonolite Pròcida: K-trachybasalt-latite
Campanian Ignimbrite	34 000	100	K-phonolite
Neapolitan Yellow Tuff	12 000	20	Trachyte–phonolite
Gauro	12 000–8000	0.7	K-phonolite
Agnano-1 (Pomici Principali)		0.5	K-trachyte
Archiaverno		0.3	?
Minopoli		0.01	K-trachybasalt
Agnano-2 (Montagna Spaccata)		0.15	K-trachyte
Concola–Fondo Riccio		0.01	K-trachybasalt–K-phonolite
Nisida		0.08	K-phonolite
	4500–(?)3000		
Agnano-3 (Mt Spina-Pigna S Nicola)		0.08	K-trachyte–K-phonolite
Capo Miseno, Baia, Fondi di Baia		0.4	K-phonolite
Cigliano		0.1	K-trachyte
Olibano		0.08	K-phonolite
Solfatara		0.1	?
Astroni		0.5	K-trachyte–K-phonolite
Averno	?		K-phonolite
Senga		0.06	K-phonolite
Monte Nuovo	~500 (AD 1538)	0.02	K-phonolite

It contains three marine horizons, each marking an oscillation about sea level, the last uplift occurring 4500 years ago. Since then, eruptions have been concentrated around the central coastal zone of Campi Flegrei, close to the rim of the NYT caldera and virtually all within 3 km from Pozzuoli. The last eruption, in 1538, occurred after a possible hiatus of 2500 years and produced the modest Monte Nuovo (20 million m³), along the coast between Pozzuoli and Baia. Recent unrest has taken the form of slow ground oscillations (about 1 cm per year) around Pozzuoli, occasionally punctuated by rapid uplift at rates of decimetres per year.

Apart from the Averno and Agnano complexes, most eruptions have occurred from monogenetic vents, each producing cones typically 1 km or less across and about 100 m high (the Gauro cone is the tallest at 336 m). As a result, Campi Flegrei appears from above as a region pockmarked by craters, a feature used in the late nineteenth century to support a volcanic origin for the craters on the Moon (Fig. 3.1).

As might be expected, the geological composition of Campi Flegrei is dominated by pyroclastic materials, of which the majority are flow (surges, flows and ignimbrites) rather than fall deposits. Most of the units more than about 10 000 years old have become lithified as *tuffs* and have been used extensively for construction (from building palaces to paving roads). Indeed, it was by mixing powdered tuff with water that the ancient Romans rediscovered cement.

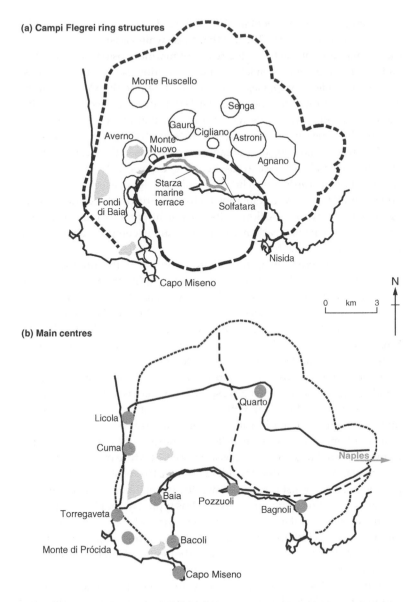

(a) Campi Flegrei ring structures

Monte Ruscello

Senga

Gauro   Cigliano

Averno   Monte   Astroni
         Nuovo

                    Agnano

Starza
marine
terrace   Solfatara
Fondi
di Baia

                    Nisida

Capo Miseno

N

0   km   3

(b) Main centres

Quarto

Licola

Cuma

                    Naples

         Baia
Torregaveta   Pozzuoli   Bagnoli

                    Bacoli
Monte di Prócida

         Capo Miseno

Figure 3.1 **(a)** Campi Flegrei consists of two ring structures between which many pyroclastic cones have been formed in the past 12 000 years. The inner ring marks the rim of the NYT caldera. The outer ring denotes the *apparent* rim of the CI caldera, modified by marine erosion. Uplift along the southern coast has produced the Starza marine terrace. **(b)** The main centres are serviced from Naples by the Cumana (through Pozzuoli) and Circumflegrea (via Quarto) railways (continuous lines) and, until Pozzuoli, by the Metropolitana railway, which utilizes the national overground route that turns north past Quarto (dashed line). Intermittent services (not shown) run between Licola and Torregaveta.

Outcrops of the Neapolitan Yellow Tuff dominate exposures in the walls of the outer caldera. The fact that these walls consist of NYT and not Campanian Ignimbrite (CI) is the result of marine erosion. After the NYT had filled-in the depression left by the CI eruption, several marine transgressions flooded Campi Flegrei, the sea cutting into the NYT to form the current caldera margins. The NYT consists of K-trachytic and K-phonolitic pumice and lithic fragments within a matrix that can vary from fine to coarse ash. Appearing massive from a distance, NYT outcrops often show centimetre-scale layering (attributable to variations in the mean grain size of the matrix) and metre-scale cross bedding. Occasionally, the rock can be seen grading upwards from a lithified yellow facies to a less coherent and greyer unit. The yellow colour appears to be the result of secondary zeolitization as the pyroclastic flow units became lithified to a tuff. The lithic fragments, which may reach metres across, vary in composition from grey trachytic lava to green tuff.

Non-NYT outcrops in Campi Flegrei consist of the products from relatively small, local eruptions. Material pre-dating the NYT is best seen in the walls of the caldera (especially near Camaldoli and Soccavo), along the southwest coast of Campi Flegrei (especially Monte di Pròcida) and on the adjacent island of Pròcida. Most distinctive is the Piperno–Breccia Museo formation. The Breccia Museo ("museum breccia") is a poorly sorted coarse-grain body containing heterogeneous lithic breccia (typically less than 1 m across) in a matrix of glassy scoria. The lithics represent a wide range of the underlying crust, including lavas, tuffs and sedimentary rock, and it is this variety that gives the unit its name. The Piperno unit, best seen in the northeastern caldera wall near Soccavo and Camaldoli, and famous for its use in buildings, consists of flattened elongated pumice sintered together – the classic eutaxitic texture of welded tuff.

Pyroclastic cones across the caldera floor are the most-evident post-NYT products. They consist mainly of trachytic (sometimes alkaline) pumice-fall units and pyroclastic surges. With rims typically about 100 m tall, the widths of single craters range from some 100 m to 2 km. The resulting ratios of crater width to cone height (W_{CR}/H_{CO}) thus range from about 1 to 10, a variation that can be attributed to the degree of interaction between magma and groundwater during eruption.

In the simplest case, when groundwater interaction is not important and magma is ejected ballistically from explosions within a narrow conduit, the cones around vents tend to build themselves up with a mean internal angle of 45°, so that values of W_{CR}/H_{CO} cluster around 2. However, when groundwater does mix significantly with fragmenting magma in the throat of a volcano, it cools the magmatic mixture while flashing to steam. The two effects counterbalance each other energetically. For modest proportions of added water (about 10–20 per cent by weight according to laboratory experiments), the effect of additional pressurized steam dominates, encouraging both further fragmentation and higher exit velocities for the magma. This combination favours the formation of wet pyroclastic surges that can travel distances greater than those expected from ballistic trajectories, producing cones with values of W_{CR}/H_{CO} of 10 or more. However, when much larger amounts of water are added, the cooling effect dominates and the pyroclastic surges become increasingly sluggish. As a result, the associated values of W_{CR}/H_{CO} again decline below 10. As shown in Figure 3.2, the cones in Campi Flegrei provide excellent examples of the complete range of geometries for hydromagmatic cones, from the near-magmatic (Monte Nuovo), through the energetic (Averno) to sluggish (Capo Miseno) hydromagmatic products.

Completing the cast of volcanic products in Campi Flegrei are scattered domes of K-phonolitic lava. Although volumetrically they have a minor role, domes have been emplaced throughout the volcanic evolution of the area. The earliest domes occur close to and beyond the outer edges of the caldera, at Punta Marmolite, near Quarto (47 000 years old) and at Cuma (37 000 years old), both of which pre-date the Campanian Ignimbrite. At the other extreme, examples younger than about 4000 years can be found

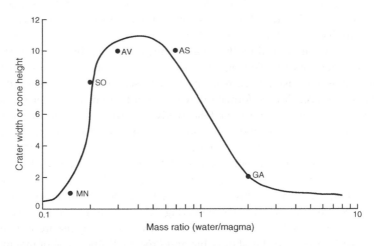

Figure 3.2 Changes in the ratio of crater width to cone height (W_{CR}/H_{CO}) among hydromagmatic cones. W_{CR}/H_{CO} is a crude measure of the efficiency with which vaporizing water can propel erupting magma. The curve is indicative only, to emphasize the expected peak in W_{CR}/H_{CO} with changes in the mass ratio of interacting groundwater and magma. Note the logarithmic scale for the water:magma ratio. The labels for the craters are given in Figure 3.1.

within the central-eastern part of the caldera, along the Starza marine terrace and beneath the crater walls of Solfatara (the Monte Olibano complex)
and within Astroni, 2 km to the north. In the case of Astroni, which boasts
three domes, the lava represents the final effusive stage of an initially
explosive eruption. The other domes may also have been late-stage eruptive products or, instead, the result of independent effusive events.

Campi Flegrei and the Campanian Ignimbrite

Although the caldera produced by the NYT eruption has been clearly identified in the Bay of Pozzuoli, the association between Campi Flegrei and
the Campanian Ignimbrite is less straightforward (Fig. 3.3). The standard
interpretation has the CI erupting from ring vents (the future caldera walls
after collapse) centred upon a location just north of Pozzuoli. A recent variation on this theme has enlarged the area of caldera subsidence to include
the Bay of Pozzuoli and the district now covered by Naples.

Two problems in choosing Campi Flegrei as the vent area for the CI (as
opposed to an area that subsided after underlying magma had escaped
from vents elsewhere) result from the distribution of the deposit. The first
is that, although extensive CI deposits occur in the sector north to southeast
of Campi Flegrei, virtually nothing has been found on the islands of Pròcida and Ischià, 2 km and 7 km offshore to the southwest. Both islands
existed before the CI eruption. The lack of CI deposits would thus suggest
a remarkable directional control on an ignimbrite that, elsewhere, easily
flowed distances of several tens of kilometres. Secondly, any vent area is
expected to be marked by thick basal units containing breccia from the
rock broken apart while magma was forcing a path to the surface. Such
breccia is surprisingly scarce, apart from scattered outcrops of the
Piperno–Breccia Museo formation, which might be a distal section of the
CI base (erosion having removed the overlying CI) or the product of one or
more earlier, local eruptions. In contrast, an area of giant breccia (with
blocks metres across, larger than in the Breccia Museo) covering at least
25 km² has been traced about an axis 10–12 km long between Naples and
Villa Literno to the northwest, looking suspiciously like an extension of the
almost linear (and possibly fault-bounded) coastline running southeast
from Naples past Somma–Vesuvius (Fig. 3.3). An alternative interpretation, therefore, is that the CI was fed from a NW–SE trending fissure system,
at least 10–15 km northeast of Ischià and Pròcida. Withdrawal of magma
induced regional subsidence, perhaps defining the present Bay of Pozzuoli and Gulf of Naples, as well as forming the original Campi Flegrei
depression without the need to invoke collapse along active ring vents.

Figure 3.3 Distribution of the proximal deposits from the Campanian Ignimbrite. The inland edges are controlled by topographic relief. The zone of giant breccia may define the source fissure. This fissure appears as an extension of the almost linear coastline by Somma–Vesuvius, suggesting a location controlled by tectonic faults. The distribution assumes that the Breccia Museo in Campi Flegrei is a product of local eruptions and not an early phase of the CI. An alternative interpretation, still consistent with the postulated source fissure, is that the Breccia Museo represents the distal portion of the CI basal breccia.

Campanian Ignimbrite   giant breccia

The eruption of Monte Nuovo, 1538

In sixteenth-century Campania, volcanic activity was rarely a topic of conversation. Somma–Vesuvius had been quiet following a modest outburst in 1139 and, since then, the only surprise had been the 1302 Arso eruption on Ischià. The real menace came from earthquakes, even low-intensity events (about III on the Mercalli scale) producing cracks in buildings. Indeed, the only curiosity in Campi Flegrei had been between 1501 and 1503, when a gradual extension of the beach near Pozzuoli caused rivalries to flare up as to who owned the new land. No-one imagined that uplift of the new beach might be a precursor to an eruption in less than 40 years' time.

More ominous omens arrived in 1536, a year when central Campi Flegrei was shaken by intermittent trains of small earthquakes. Virtually continuous by the summer of 1538, the degree of ground shaking dramatically increased on 27 September, especially between Pozzuoli and Lago Averno, where uplift had forced the sea suddenly to withdraw. At 01.00 h on 29 September, cold water poured from cracks in the swollen ground near the village of Tripergole. Steam and mud soon followed until the rising magma, mixing with groundwater, burst out in a violent hydromagmatic eruption that sent wet surges flattening buildings as far as Pozzuoli, 3 km to the east. Within 48 hours, Tripergole had been buried by a new cone (Monte Nuovo), 133 m high and 700 m across.

With the growth of the cone, activity declined for the next four days to long pauses interrupted by occasional bursts of Strombolian explosions. The decrease in eruptive violence encouraged bystanders on 6 October to climb the cone along a small depression in its coastal flank. The timing was unfortunate. Before the crater rim had been reached, flows of incandescent scoria swept down the trough, engulfing the adventurers and killing at least 24 people. It was the last, tragic act that brought the eruption to a close.

Involving some 20 million m^3 of magma, the Monte Nuovo eruption was one of the smallest on record in Campi Flegrei. Even so, the event has proved crucial to understanding eruptive behaviour in the district.

First, the initial outpouring of water and early hydromagmatic explosivity both confirm the importance of groundwater in controlling the style of eruption. Indeed, the subsequent change from hydromagmatic to Strombolian activity is consistent with a decrease during eruption in the proportion of groundwater that could filter into the magmatic system. Secondly, accounts of buildings destroyed as far as Pozzuoli highlight the danger from even modest hydromagmatic pyroclastic surges. Deposits from these surges can no longer be found in the field. Without the eyewitness records, the presence of early surges in hydromagmatic events would have passed unrecognized, resulting in unrepresentative hazard maps for Campi Flegrei. Thirdly, and finally, the Monte Nuovo eruption was preceded both by coastal uplift nearly forty years beforehand and, just before it started, by almost two years of persistent low-intensity seismicity. Since 1969, a new crustal bulge has been developing around Pozzuoli, together with persistent, low-intensity seismicity between 1983 and 1985. A nagging doubt thus remains as to whether or not Campi Flegrei is again gradually building up to its next outburst.

The next eruption in Campi Flegrei

Since the NYT eruption, at least what is now the southern coast of Campi Flegrei has been subject to slow oscillations about sea level. Evidence comes from the Starza marine terrace, from the presently submerged Roman ruins running offshore from Baia to Pozzuoli, from observations of the Roman Serapeo in Pozzuoli (itinerary below), and from two episodes of rapid uplift in 1969–72 and 1983–5.

The background oscillations or bradyseisms (Greek: "slow movement") have had an average velocity of centimetres per year, similar to the rates of movement along tectonic plates (or the growth rate of fingernails). Although the bradyseism might be connected to tectonic movements in the mantle, a simpler explanation is that they are attributable to the oscillating rebound of the crust after its collapse into the partially drained magma reservoir that fed the NYT.

Indeed, most of the slow movements have occurred close to the northern rim of the NYT caldera, yielding a net uplift on the northern side (typified by La Starza) and a net sinking just offshore to the south (shown by the submerged Roman ruins). Such a situation is consistent with the long-term subsidence of the NYT caldera roof squeezing underlying magma upwards to the north, thereby also explaining the predominance of post-NYT activity in the central-southern part of onshore Campi Flegrei.

Another feature of post NYT-activity is an apparent inward migration of eruptive centres towards the Pozzuoli district. In the period 12 000–8000 years ago, eruptions occurred from Baia, through Gauro and Agnano to Nisida, in an irregular arc around Pozzuoli with a mean radius of about 5 km. In comparison, eruptions after the hiatus ending 4500–4000 years ago have occurred at distances less than 3 km from the town, and it is within this area, on the landward (northern) uplifting part of the oscillating caldera that a future eruption seems most likely to take place.

Further evidence for this scenario comes from two episodes of rapid uplift that have shaken the Pozzuoli district since 1969. These episodes, in 1969–72 and 1983–5, involved uplift of a bulge some 4–5 km in radius centred upon Pozzuoli. In both cases the amount of uplift increased markedly within a 3-km radius, reaching peak values beneath Pozzuoli of 1.5 m and 1.8 m in the 1970s and 1980s respectively. Allowing for minor settling, Pozzuoli today stands 2.3 m higher above sea level than it did in early 1969. The first episode was virtually aseismic, but, even so, the threat to building collapse caused evacuation of Pozzuoli's historic centre. The second episode was accompanied by a high level of persistent ground vibration, ranging from seismic swarms (hundreds of small tremors) lasting several hours to larger earthquakes reaching magnitude 4 on the Richter scale. Thirty thousand people were evacuated, some moving away, but most electing to live in tents and makeshift cabins along the shore.

At decimetres per year, the rates of uplift were ten times faster than the typical bradyseismic movements, an increase that can be explained most easily by a batch of magma approaching the surface. Analyses of the deformation suggest that the magma batch is a WNW-trending dyke 3 km long and about 3 km deep, the centre of which lies a few hundred metres east of Pozzuoli – virtually an eastward extension of the Starza marine terrace.

The appearance of seismicity during the 1983–5 uplift shows that the cumulative strain on the bulge had exceeded the value necessary to induce significant crustal cracking. If this strain is not relieved by mass re-adjustments (not necessarily deflation), then another rapid uplift can only bring closer the possibility of increased fracturing and the eruption of magma.

Excursion time. Three to four hours on foot, excluding travel time to and from Arco Felice. The excursion can be combined with a visit to Pozzuoli and Solfatara (next itinerary).

Public transport. Monte Nuovo is about 15 min walk from the Arco Felice Cumana station. Local bus services also run to and from Naples and from other towns in Campi Flegrei.

Maps. Istituto Geografico Militare topographic maps: 1:25 000 Sheets 184/III/ NE, 184/IV/SO. Touring Club Italiano 1:50 000 topographic map, II Golfo di Napoli, 2. Consiglio Nazionale delle Ricerche, 1:25 000 geological map of Campi Flegrei. As for Somma–Vesuvius, these maps may be difficult to obtain.

Difficulty. Easy to moderate. Monte Nuovo can be climbed slowly along a dirt track in about 15–20 min. The path into the crater is steeper.

Hydromagmatism in Campi Flegrei

Monte Nuovo is the product of the only recorded eruption in Campi Flegrei. It is a well preserved, mixed Strombolian–hydromagmatic cone and is believed to be the best example of what a future eruption will produce. The western side of Monte Nuovo overlaps the Averno cone which, together with Astroni, shows the extreme morphology of strongly hydromagmatic cones (Fig. 3.2).

From the Cumana station at Arco Felice, turn left past the level crossing and follow the road 500 m to the first crossroads. Turn left (west) into the shopping street that forms part of the main coastal road between Pozzuoli (east) to Baia (west). After about 500 m, take the side road that opens diagonally up hill to the right, by a brown sign indicating the "Oasi Naturalistica di Montenuovo".

The side road is built over the lower flanks of Monte Nuovo. Gaps in the vegetation on the right allow the first glimpses of a fine-grain pale brown tuff overlain by deposits of grey-black angular fragments decimetres across. After 250 m, the road gives way to a grassy terrace (where wooden tables and chairs have been set up for picnics) and to the entrance to the protected area of Monte Nuovo, newly established by the local council and run by volunteers. The protected area is open to the public on Monday to Friday from 08.30 h to one hour before sunset, and on Saturday, Sunday and public holidays from 08.00 h to 13.00 h. Entrance is free and guided tours can be arranged (081-804.1462); advance warning of large groups (about 15 or more) is appreciated.

The path to the crater rim begins with some steps to the right of the custodian's office at the entrance to the protected area. Before ascending, it is worth looking at the extensive outcrop that forms the backwall of the terrace. The terrace cuts across the cone's south-flank depression, along which visitors were killed near the end of the 1538 eruption. The backwall consists of dark-grey, fragmented and poorly vesicular scoria. Mostly decimetres across, the angular fragments dominate the matrix that grades from centimetre-size debris to dust (Fig. 3.5). About a metre from the top, a subhorizontal layer of fine-grain material stands out in relief, suggesting that the whole deposit may have been emplaced in two stages, the interval being marked by the settling of fine particles from a volcanic cloud.

The scoria deposit tapers out after about 100 m, marking where it was contained by the western margin of the depression (Fig. 3.5). The obvious

Figure 3.5 The Monte Nuovo scoria flow pinching out (right to left) over the yellow pumice (low, centre) forming the bulk of the volcanic cone.

topographic control demonstrates that the deposit was emplaced as a flow. Unless the scoria was delivered by a lava fountain tilted towards the depression (or by directed Strombolian explosions), it seems likely that the unit was the product of rim collapse. Thus, the shape of the crater rim allowed scoria from lava fountains to collect within the depression, constructing an unstable wall. The wall collapsed, feeding the first flow. Continued fountaining enabled a second wall to develop and the process was repeated.

The bulk of the cone consists of yellow-white pumice, dominated by angular pieces centimetres or less across. Much of the colouring is super-ficial and the pumice fragments retain a pale grey interior. Although at first glance homogeneous, the unit in fact shows a weak layering defined by whether or not a layer contains centimetre-size pumice. Some large rounded pumices, decimetres across, occur throughout the outcrop, occa-sionally associated with impact sags in the finer material. Poorly vesicular angular grey fragments, centimetres across, are also scattered throughout the deposit.

The angular nature of small pumices suggests that they are the shat-tered remains of larger parents. Shattering may have occurred during em-placement; however, it may also have been provoked by steam explosions, as pumice reaching the throat of the volcano vaporized groundwater. Indeed, the angular grey fragments may be the shattered component of the least-vesiculated fraction of ascending magma.

The contrast in colour and form between the pumiceous units and the scoria flow is all the more striking since both consist of K-phonolite and are chemically indistinguishable. Their differences reflect variations in phys-ical conditions upon eruption, presumably a result of earlier magma having been more vesicular and more interactive with groundwater.

From the outcrop, return along the terrace to the steps leading up the cone. On a clear day, the terrace offers a good panorama of western Campi Flegrei, from Baia (note the castle), Capo Miseno and Ischià on the right, to Pozzuoli's old town and promontory (and, possibly, a shadowy Capri in the distance) to the left.

The steps end in a dirt track leading up the tree-lined cone. After about ten minutes, the track reaches a walled path. Turn left up the path, which shortly opens onto the summit rim. The first view of the crater reveals a heavily vegetated and steep-sided inner wall, dotted by landslide scars. The wall drops some 120 m to the flat crater floor (14 m above sea level), 100–150 m across. Four pillars at the centre of the floor surround an explor-atory geothermal borehole. The crater floor can be reached along a narrow track leading from behind the end of the inner stone wall flanking the path to the summit. The path cuts through the tuff forming the cone, although good outcrops are rare because of the vegetation. More rewarding is a walk around the rim of the cone, starting by continuing straight ahead from the end of the walled path.

Although an impediment to geological studies, the heavy vegetation is a testament to the fertile nature of the volcanic products. Along the path itself, occasional boulders of dense pumice reveal the violence of activity even towards the end of the cone-building stage of the eruption. After about 350 m, the path divides at the corner of a ruined building. Turn left along the branch along the outer rim of the cone, where several excellent viewpoints can be found on the left, looking over Lago Averno.

Lago Averno (Fig. 3.6) fills the inner of two craters that make-up the Averno complex. Formed by an eruption about 3700 years ago, the inner crater is nested within the older cone of Archiaverno ("ancient Averno"). Both cones consist of ballistic pumice and pyroclastic flows. The craters are notably very wide (about 2 km) with respect to their heights (about 200 m), a feature characteristic of strong hydromagmatic activity.

Averno takes its name from the Greek for "without birds". At least 2100 years ago, the crater lake was actively degassing, the gases collecting amid the then thick tree cover of the crater's inner walls. Birds settling among the trees rarely escaped alive (from descriptions, it seems that CO_2, which is denser than air and collects in hollows, was the primary killer; recall the deaths caused by CO_2 during Vesuvius' 1944 eruption in Ch. 2). For this reason, Virgil associated Averno with the entrance to Hades.

During one of Rome's many civil wars, in 37 BC the naval commander Agrippa ordered construction of a canal south between Lago Averno and the Gulf of Pozzuoli, so that the lake could be used by his fleet. The remains of the coastal port now lie submerged offshore, a result of bradyseismic displacements. Degassing also appears to have declined at about this time. Whether or not due to Roman engineering works, the end of degassing encouraged the later construction not only of the Temple of Apollo (which may in fact have been thermal baths) on the eastern shores of the lake, but also of a tunnel through the northwest rim of the cone, as part of another canal leading to the west coast of Campi Flegrei. The tunnel is today known as one of the Sybil's caves (Pseudogrotta della Sibilla) and can be visited by appointment (081-526.7256). Remnant heat flow from the area can be enjoyed in the natural sauna, the Stufe di Nerone, along the coastal road just south of Averno, towards Baia.

The Temple of Apollo can be seen in the foreground, below the rim path on Monte Nuovo. The building was surrounded by some of the earliest pyroclastic surges from the 1538 eruption. However, unlike the eastward surges that destroyed houses on the outskirts of Pozzuoli, the surges moving northwest were too diluted by volcanic gas to cause significant

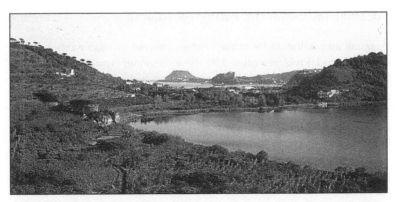

Figure 3.6 (Opposite) Lago Averno as seen in the late eighteenth century, from William Hamilton's *Campi Phlegraei* (1776) and **(Above)** today. The views are looking south, along the artificial entrance dug by the Romans (the water of Lago Lucrino can just be seen in the earlier picture). The overlapping flank of Monte Nuovo (left) sweeps down behind the ruins of the Temple of Apollo. The background shows the southwestern limb of Campi Flegrei, including the flat-topped Capo Miseno and, to the right, the Aragonese castle of Baia.

damage. To the left, beyond the artificial exit to the sea, it is just possible to glimpse the water of Lago Lucrino. This lake derives its name from lucrum ("lucre"), reflecting the profit to Roman entrepreneurs who used it for cultivating freshwater oysters.

Following the gentle descent of the path for about 100 m, take the first fork to the right and return to the rim of Monte Nuovo's cone. Straight ahead (northeast), the twin peaks of Monte Gauro (or Monte Bárbaro) to the right and Monte Corvara to the left dominate the skyline. These peaks form part of the largest cone in Campi Flegrei (reaching 336 m above sea level). The western flank (to the left) has been heavily eroded and landslide scars can be seen cutting its oversteepened slopes. From here, either turn right and return along the rim to the walled path or, for general views of the landward side of Campi Flegrei, turn left and follow the path around the whole rim (note that, for the second route, the final descent is steep back to the walled path).

Return to Arco Felice along the route used for the ascent. For a closer look at Lago Lucrino and Lago Averno, either turn right at the bottom of the side road leading from Monte Nuovo and continue west for another 1.5 km, or return to the Cumana railway and take the train to Lucrino station. There, advantage could also be taken of the Stufe di Nerone (500 m farther west, beyond the entrance to Lago Averno, and on the inland side of the coast road) a rare chance to have a sauna (27–48°C) inside a volcano.

Pozzuoli and Solfatara: riding the caldera

Pozzuoli and Solfatara lie at the centre of recent ground movements in Campi Flegrei. Problems caused by these movements and how the population have responded are well illustrated by a visit along the old and new ports at Pozzuoli – conveniently also the gastronomic capital of the district. Solfatara is the type locality for solfataric processes and, in part transformed into a camping site, again highlights the uneasy relationship between human and volcanic activity.

The excursion starts by the Roman Serapeo, just inland from the tourist port and about 100 m west of the Cumana station at Pozzuoli. From here, it continues around the port and then back through the town before climbing up to Solfatara.

Some 65 m square, the Serapeo is dominated by three columns, 12.5 m tall and each sculpted from a single piece of marble (Fig. 3.7). Unlikely as it may seem, the columns are among the most famous symbols in the history of modern geology: from an Anglo-Saxon perspective, they were extensively studied by Charles Babbage, who opened the way to modern computing, and formed the frontispiece in Charles Lyell's *Principles of geology*. The columns are also responsible for the early misinterpretation of the ruins as an ancient temple. In fact the whole square seems to have been a marketplace, the bricked-off areas along the walls marking the foundations of individual shops (although the inland end of the ruins, next to the columns, may have served as a shrine to the god Jupiter).

An obvious oddity of the columns is that, 3.6 m above the ground, each contains a pockmarked band about 2.7 m thick (Fig. 3.7). The scars are the result of boring marine bivalves (*Lithodomus lithophagus*) that thrive close below sea level. Short of an attack of flying clams, therefore, these horizons indicate that the columns have sunk into and re-emerged from the sea at least once since their construction.

Excursion time. Four hours on foot, excluding travel time to and from Pozzuoli. The excursion can easily be divided into two – visiting Pozzuoli and Solfatara on separate occasions – or combined with a visit to nearby Monte Nuovo (previous itinerary).

Public transport. Pozzuoli is serviced by the Cumana and Metropolitana railway companies. The Cumana station is more convenient for the port; the Metropolitana station is better for Solfatara. Local bus services also run to and from Naples and other towns in Campi Flegrei.

Maps. As for previous itinerary.

Difficulty. Easy. Virtually the whole excursion is along paved road. Although diffuse, gas emissions from Solfatara may irritate those with respiratory difficulties.

Figure 3.7 The Roman marketplace, Il Serapeo, at Pozzuoli, looking south towards the coast. The dark rings around the lower parts of each column are attributable to bivalves boring into the marble. The bases of the columns were protected by accumulations of sediment and volcanic ash.

Initial submergence of the columns must have occurred after the first half of the third century AD, since records exist of renovations to the area as late as AD 235 and the perforations attributable to the bivalves post-date any restoration (none of the perforations contain traces of the cement expected if the Romans had tried to cover the pockmarks). The bases of the columns re-emerged during the early 1500s, during the period of local

75

uplift that preceded the eruption of Monte Nuovo (see previous excursion), and excavations in 1750 showed that, while submerged, the columns had been surrounded by sequences of volcanic products and marine sediments, material that had protected the lower portions of the columns from bivalve activity.

By the start of the nineteenth century, the floor of the Serapeo was again being inundated with sea water during high tide and continued to submerge for about the next 150 years at an average rate of 14 mm per year. Sinking halted abruptly in 1968, when the whole Pozzuoli district began suddenly to rise and, in just three years, carried the Serapeo upwards 1.5 m, almost to its level in 1820. Following a slight subsidence of 0.25 m between 1972 and 1982, a new 30-month phase of uplift raised Pozzuoli another 1.8 m out of the sea. By the end of 1986, the Serapeo had subsided slightly by about 0.27 m.

Cross to the coast road on the far side of the Serapeo and turn left (towards the town centre). During the 1982–4 uplift, the tree-lined squares to the left (as well as that on the coastal side of the Serapeo) were filled with emergency shelters that evacuees had transformed into a variety of shops. At the height of the emergency, it was almost surreal to see a barber shaving the foam-covered chins of his customers out in the open.

After 150 m, turn right to the ferry port. If you are lucky, the fish market will add a touch of colour and confusion. Notice on the right that the port has been built on two levels. This is a measure of the 1.8 m uplift during 1982–4: the upper level was by the sea before 1982 and a new, lower level had to be built for mooring small boats and to allow the ramps from car ferries to incline at safe angles from ship to shore. Looking west across the bay, note the tree-lined cone of Monte Nuovo and recall that pyroclastic surges from the 1538 eruption almost reached the location of today's port.

At the far end of the port, follow along the coast to the left until reaching the old fishing port with, behind, the *Rione Terra*, the historic centre of Pozzuoli. The Rione itself was evacuated during the 1969–72 bradyseismic emergency and has yet to be reoccupied. The decision to evacuate remains controversial, some still convinced that, in addition to the emergency, there was a hidden agenda to transform the Rione into a luxury tourist complex. More recent projects include the possibility of reopening the district as an archaeological museum. Before 1982, the fishing port was a crowded haven for small fishing boats. However, uplift rendered mooring extremely difficult and also reduced the depth of water in the port by nearly 2 m. It is still used by some boats, but not at the level of 20 years ago.

Continue beyond the port along the narrow street with new restaurants to the left and the Rione to the right. Turn left at the end of the street and enter the small square. Turn right and continue up hill out of the square, passing beneath a bridge where the road turns sharp right. Take the road on the left immediately after the bridge and wind up hill along via Marconi and the connecting via Rosini. After about 500–600 m, via Rosini merges with via Solfatara at a triple junction. Pozzuoli's Roman amphitheatre (l'anfiteatro Neroniano-Flavio, the third largest in Italy and open from 09.00 to one hour before sunset; 081-526.6007) is on the opposite side of the road. With the amphitheatre on the left, continue up hill, passing beneath the Metropolitana railway bridge (station 100 m along the road to the left (northwest) on the uphill side of the bridge) and follow the road for another kilometre until arriving at the entrance to Solfatara along a side road to the left (keep to the main road and ignore faded signs suggesting a shortcut to Solfatara).

Before visiting the crater (open 08.30–19.00 h), it is first worth climbing the steep road passing to the left of the entrance (the road is narrow and without adequate parking, so if in a vehicle, it is better to park first and then to continue on foot). After 300 m, the road affords an excellent over-view to the right of the whole Solfatara complex (Fig. 3.8).

About 3500–4000 years old, Solfatara is a wide flat-bottomed crater, some 650 m across, bounded by irregular steep-sided walls 80–85 m tall. A depression to the southwest (the current entrance) is the probable scar of partial cone collapse. The western part of the cone, beneath the road, consists entirely of pyroclastic flows overlying a basal breccia; around the rest of the cone, the breccia and pyroclastic material lie on top of earlier K-phonolitic lava domes, including Monte Olibano, a part of which can also be seen at the top of the eastern end of the Starza raised marine terrace.

Solfatara is the most celebrated zone of fumarolic activity in Campi Flegrei. Weathering and mineral deposition have weakened the crater walls (encouraging enlargement of the original crater through collapse) and coated them with a spectrum of white and yellow hues. Tree planting has created a green oasis in the southern part of the crater, where it is now a popular camping site and even boasts a swimming pool. In contrast, the northern half of the crater maintains its savage appearance. Steam from fumaroles can be seen along the northeastern and, to a lesser extent, the western walls. Persistent degassing made Solfatara a natural target for mining and during the late eighteeth century the volcano yielded 27 300 kg of sulphur, 3700 kg of alum and 200 kg of ammonium salts annually. Today its fumaroles are continually monitored for any chemical changes (mainly an increase in the proportion of SO_2) that might presage the arrival of magma at shallow depth beneath the Pozzuoli area.

Return to the main entrance of the crater, which is privately owned and can be visited by payment only (note that payment also provides access to a car-park within the crater). Along the path beyond the ticket office,

Figure 3.8 The eastern interior of Solfatara crater **(top)** reveals a desolate landscape of ash and pumice coated by sulphur-rich deposits. The main fumarole field behind the Freidlaender building **(bottom and top, left)** includes the Bocca Grande, from which gases escape at temperatures of as much as 160°C.

follow the signs marked "Al Cratere" ("to the crater"). Emerging from the trees, the unvegetated part of the cone at first seems blindingly bright, thanks to the ubiquitous cover of pale mineral deposits. Continue right along the fenced-off central part of the crater. Until 1983, mudpools bubbled almost continuously in the middle of the crater floor. These were drained when a new fracture opened during the bradyseismic activity of 1984. Today, only ephemeral pools occur after heavy rainfall.

Moving away from the centre, the path continues to the main fumarole field along the foot of the northeastern crater walls. Even though they are fenced off, care should be taken near the fumaroles since the ground is liable to collapse (a new high-pressure fumarole opened in the field in December 1984). The lonely stone building (Fig. 3.8) was established by Immanuel Friedlaender, who around the turn of the twentieth century dedicated himself to studying Solfatara's emissions, especially those from the so-called Bocca Grande ("large mouth"), just behind the building, where temperatures have been reported as high as 160°C, compared to the 90–99°C common for most of the fumaroles.

Follow the crater wall westwards to a double-arched structure set into the cliff face. Steam collecting in these arches is reputed to be invigorating. One test is to stay in an arch long enough to have a photograph taken. Otherwise, take the opportunity to explore the highly altered face of the crater wall, noting the greater resistance of dense lithics compared with the enclosing pumice.

Leaving Solfatara, a good panorama of the bay can be had immediately opposite the side road leading to the crater. Turn right and follow the road down hill to Pozzuoli, perhaps to take advantage of one of the excellent seafood restaurants along the port or between the port and the Serapeo.

Other sites to visit

Agnano

Rather than a place to visit, Agnano is a site to ponder while passing by. The product of three major episodes of activity (Table 3.1), Agnano is a multiple crater, 3 km wide, which contained a lake until it was drained in 1870. Since then the area has been overcome by urban sprawl and today hosts a hippodrome famous among horse-trotting circles. Agnano also boasted thermal spas during Roman times and one of these, the Stufe di San Germano, still operates near the trachytic dykes of Monte Spina ("thorn mountain") in the southwest corner of the complex (and which can be reached from the Cumana station Agnano; 081-570.2122). Until the start of the twentieth century, a grisly heritage of the degassing was a guided tour to the Grotta del Cane ("dog's cave") in the southern crater wall. As a result of gas emissions, the lowermost 30–40 cm of the cave were persistently covered by carbon dioxide. Visitors were shown the effects of the gas as stray dogs were led into the cavern where they were asphyxiated.

Astroni

Today a reserve run by the World Wildlife Fund, Astroni is with Averno a classic example of a strongly hydromagmatic cone. Its crater, 2 km across, contains three lakes, as well as a scoria cone, Colle dell'Imperatrice ("empress's hill"), and K-phonolitic lava domes, Rotondella ("small rotunda") and Paglieroni ("straw fences"), which, along with K-phonolitic intrusions into the eastern crater wall (Caprara) and northwestern ridge of the largest lake, Lago Grande, are the final products of Astroni's activity. Used as a spa by the Romans, Astroni was for nearly 300 years a royal hunting reserve, until King Ferdinand II opened the cone as a public park in 1830. Since 1987, the crater has been declared a World Wildlife Fund reserve protecting local plants and animals. As a result, vegetation is rife within the crater and, although scenic and tranquil, this has covered many geological outcrops. However, along the path from the entrance to the crater floor it is still possible to glimpse sections of cross-bedded pyro-clastic surges containing lithic fragments of yellow tuff and grey trachytic lava. The crater can be reached by bus (ATAM bus C12) from the Metro-politana station Napoli Campi Flegrei. Currently, it is open to the public at weekends from 10.00–14.00 h (groups by appointment) and to educa-tional parties by appointment on Mondays, Wednesdays and Thursdays. Greater access may be available in the future, and the latest information can be found from the Reserve on 081-588.3720.

Cuma

A K-phonolitic lava containing the mineral sodalite, the Cuma dome is best known for the rediscovery in 1932 of the ancient Sibyl's cavern, a quadrangular chamber reached via a carved tunnel 115 m long and 18 m high. The outer surface shows some flow patterns, mostly highlighted by trains of vesicles. Elsewhere, vesicles have accumulated to form giant bub-bles decimetres across. Banking on one side of the dome are remnants of the Neapolitan Yellow Tuff. Cuma can be reached along the Circumflegrea railway (six trains a day) that runs from Naples, via Quarto to western Campi Flegrei. Other Circumflegrea trains (the great majority) stop at Licola, about 2 km north of Cuma.

Torregaveta and Acquamorta

Beach sections along the southwest tip of Campi Flegrei offer excellent opportunities to see the Neapolitan Yellow Tuff and older products. Prin-cipal sections are at Torregaveta and, 1 km farther south, at Acquamorta below Monte di Pròcida.

Just down hill from the Cumana station Torregaveta, the road leads to a beach by a small jetty. Here the Neapolitan Yellow Tuff lies unconform-ably on top of a wedge-shape sequence of tuffs and breccia from the local Torregaveta eruption and these, in turn, rest subhorizontally over the Breccia Museo.

At Acquamorta, the Neapolitan Yellow Tuff (top of section) lies unconformably on top of a series of breccias (Torregaveta deposits overlying the Breccia Museo) that rest on a thin horizon of pyroclastic flow deposits. Underneath, a sequence of alternating palaeosoils and fall deposits (from eruptions on Ischià and Pròcida) rest upon a trachytic lava dome.

Pròcida and Ischià

The volcanic islands off the southwestern tip of Campi Flegrei have good ferry and hydrofoil connections from Naples (Molo Beverello and Mergellina) and Pozzuoli. A visit to each island will entail a day trip at least.

Tiny Pròcida (3 by 1.5 km) is by far the least spoilt area in Campi Flegrei and can easily be crossed by bus or microtaxi. A good section of the Breccia Museo can be seen beyond the east end of the harbour (to the right when facing the sea). New construction is making access increasingly difficult, but not impossible with some scrambling and imagination. When clear, the same unit can be seen as a sagging deposit in the cliffs of Monte di Pròcida in mainland Campi Flegrei.

At the far (southern) end of the island (bus stop and a handful of restaurants) a small port is enclosed by the cross-bedded pyroclastic surges from the island's pre-NYT Solchiaro eruption (between 19000 and 14000 years ago). From the port, follow the road west about 50 m and, instead of continuing sharp right along the coast, turn left and up hill where the road degrades into a track. Descend the other side of the hill to a bridge connecting Pròcida to the partially collapsed cone of Vivara (it might be necessary to negotiate rusty barriers designed to keep out traffic). The remnants of Vivara itself are normally closed to the public (check locally for dates of occasional open days). Pròcida's oldest subaerial feature (about 40000 years old), the dipping beds of the Vivara cone, are overlain by the subhorizontal pyroclastic flows from Solchiaro, the youngest volcanic material from the island.

Pròcida's larger neighbour Ischià is a popular tourist destination, especially for its spa complexes. The earliest products on the island, which continues another 900 m below sea level, are about 300000 years old. Today it is dominated by its central peak Monte Epomeo (787 m above sea level), the tilted and uplifted floor of a caldera formed during eruption of the Epomeo Green Tuff (Tufo Verde del Monte Epomeo) 55000 years ago (with the uplift complete about 33000 years ago). Most of the recent activity (less than 10000 years ago) has occurred from the eastern part of the island, south of the main port, Ischià Porto, itself the remains of a crater produced in the third century BC. The last eruption in 1302 produced the Arso lava flow that destroyed the village of Geronda on its 2.7 km journey to the sea (Punta Molina, 1 km east of Ischià Porto). Although replanted with trees, the bulk form of the flow (1 km wide and 15 m thick at the coast) is evident from its topography.

Boat trips in the Bay of Pozzuoli
Ferries plying the routes from Naples to Ischià and Pròcida offer a splendid opportunity to study the large-scale morphology of Campi Flegrei and the distribution of its post-NYT cones. In particular, good views can be had of Capo Miseno, the remains of a cone which, eroded by the sea, displays the classic quaquaversal dip of layers forming volcanic cones. Capo Miseno was also the place from where Pliny the Younger watched the AD 79 eruption of Somma–Vesuvius. Worthy of note is that even at this distance from the volcano Pliny and his family were engulfed by a dilute pyroclastic cloud racing over the water.

Other trips around Campi Flegrei and the islands can be negotiated with local fishermen. Submerged Roman ruins can also be viewed from a service running glass-bottom boats from the port at Baia (information is available from the ALISEO Operations Centre in Pozzuoli's Piazza della Repubblica, 081-526.5780).

Further reading

Websites
For the Vesuvius Observatory and monitoring of Campi Flegrei:
www.osve.unina.it
For the history of Campi Flegrei and additional itineraries, including Ischià:
http://vulcan.fis.uniroma3.it
For the history of Monte Nuovo (in Italian):
http://www.tightrope.it/monten/intro.htm
For the Solfatara campsite:
http://www.kenno.com/solfatara
For a photographic tour of the Solfatara crater:
http://touritaly.org/stara/Starahom.htm

Literature
Barberi, F., F. Innocenti, L. Lirer, R. Munno, T. Pescatore, R. Santacroce 1978. The Campanian Ignimbrite: a major prehistoric eruption in the Neapolitan area (Italy). *Bulletin Volcanologique* **41**, 10–31.
Di Girolamo, P., M. R. Ghiara, L. Lirer, R. Munno, G. Rolandi, D. Stanzione 1984. Vulcanologia e petrologia dei Campi Flegrei. Bolletino della Società Geologica Italiana **103**, 349–413.
Dvorak, J. J. & G. Berrino 1991. Recent ground movement and seismic activity in Campi Flegrei, southern Italy: episodic growth of a resurgent dome. *Journal of Geophysical Research* **96**, 2309–323.
Dvorak, J. J. & G. Mastrolorenzo 1991. *The mechanisms of recent vertical crustal move-*

ments in Campi Flegrei caldera, southern Italy. Special Paper 263, Geological Society of America, Boulder, Colorado.

Giacomelli, L. & R. Scandone 1992a. *Campi Flegrei, Campania Felix: il Golfo di Napoli tra storia ed eruzioni.* Naples: Liguori.

Giacomelli L. & R. Scandone 1992b. *Campi Flegrei, Campania Felix: excursion guide to Neapolitan volcanoes* [in Italian and English]. Naples: Liguori.

Orsi, G., S. De Vita, M. De Vito 1996a. The restless, resurgent Campi Flegrei nested caldera: constraints on its evaluation and configuration. *Journal of Volcanology and Geothermal Research* **74**, 179–214.

Rosi, M., A. Sbrana, C. Principe 1983. The Phlegraean Fields: structural evolution, volcanic history and eruptive mechanisms. *Journal of Volcanology and Geothermal Research* **17**, 273–88.

Rosi, M. & A. Sbrana (eds) 1987. *Phlegraean Fields.* Volume 114, Quaderni de "La Ricerca Scientifica" , Consiglio Nazionale delle Ricerche, Rome.

Scandone, R., F. Bellucci, L. Lirer, G. Rolandi 1991. The structure of the Campanian Plain and the activity of Neapolitan volcanoes. *Journal of Volcanology and Geothermal Research* **48**, 1–31.

Scarpati, C., P. Cole, A. Perrotta 1993. The Neapolitan Yellow Tuff: a large-volume, multiphase eruption in Campi Flegrei, southern Italy. *Bulletin of Volcanology* **55**, 343–56.

Chapter 4

The Aeolian Islands

The Aeolian Islands (Isole Eolie) rise from the Tyrrhenian Sea like a chain of pearls just 15 km north of Sicily (Fig. 4.1). There are seven main islands and another half-dozen seamounts beneath the ocean surface. Together, they form a volcanic necklace, produced by the continued northwest subduction (starting in the Adriatic and dipping below the Italian Peninsula) of the African Plate beneath the Eurasian Plate (Fig. 4.1). Resting on submerged continental crust, the islands and seamounts began to develop about 1 million years ago (Table 4.1). Today, only three of the islands (Vulcano, Stromboli and Lipari) are active and one (Panarea) dormant.

Table 4.1 Dates when the Aeolian Islands emerged above sea level.

Island	Date of emergence (million years ago)
Filicudi	1.02
Salina	0.43
Stromboli	0.23
Lipari	0.16
Panarea	0.15
Vulcano	0.12
Alicudi	0.06
Sisifo (oldest seamount)	1.30*

* Below sea level.

Rock chemistry and petrography

The Aeolian magmas belong to three broad chemical families, evolving along the trends basalt–andesite–dacite–rhyolite and its potassic equivalent, as well as K-basalt–K-trachyandesite–K-trachyte–K-rhyolite. The less-evolved magmas are normally porphyritic with 25–50 per cent by volume of phenocrysts, dominated by plagioclase feldspar and clinopyroxene, together with subordinate olivine in the basaltic magmas and orthopyroxene in the andesitic magmas. The rhyolitic (and K-rhyolitic) lavas are virtually aphyric, some producing the obsidian for which the Aeolian Islands are famous.

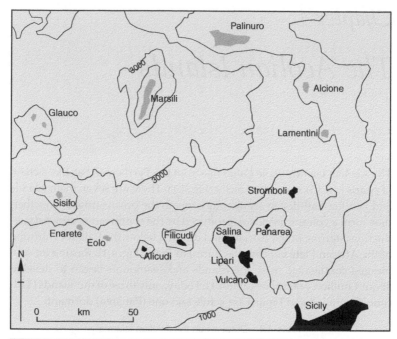

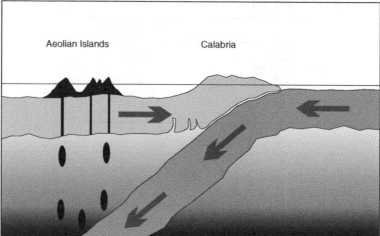

Figure 4.1 The seven Aeolian Islands (top, black) form the southeastern rim of a ring of volcanoes rising from the Tyrrhenian Abyssal Plain. The rest of the ring consists of seamounts (grey, outlines at 1000 m below sea level), dominated by Marsili and Palinuro. Bathymetric contours in metres.

The sketch tectonic section (bottom) shows the African Plate (dark grey) being subducted northwest beneath the Eurasian plate (grey). Compression where the continental crusts are colliding has produced the Calabrian Mountains of the Italian peninsula. Melting along and above the subducting plate feeds magma to the active Aeolian archipelago (black). Not to scale.

The magmas from each island tend to belong to one or two preferred suites. Basaltic–rhyolitic magmas and their potassic equivalents are dominant on the four neighbouring islands: Alicudi, Filicudi, Salina and Panarea. To the south, most of Vulcano's magmas follow the K-basalt–K-rhyolite trend (some bearing leucite phenocrysts), whereas Lipari, between Vulcano and the Alicudi–Panarea link, as well as Stromboli, to the northeast, have erupted magmas from all the chemical trends.

Vulcano

Vulcano seems unprepossessing as the island that gave its name to all locations on Earth where magma breaches the surface (as well as to a planet whose inhabitants have pointed ears). However, size is not everything and the volcano (which, in any case, continues for another kilometre below sea level) has shown a range of activity to rival that from Somma–Vesuvius.

The island is the southernmost of the Aeolian archipelago. About 8 km by 4 km, it can be divided volcanologically into five fundamental units (Fig. 4.2): South (Old) Vulcano, Piano, Lentia, Fossa and Vulcanello. These

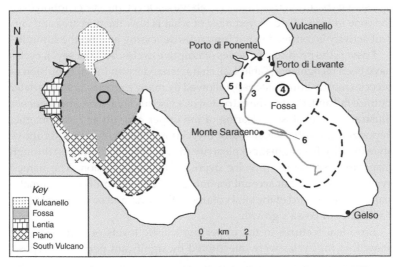

Figure 4.2 Sketch geological and location maps of Vulcano. The geological map (left) shows the five main volcanological divisions, as well as the outlines of the Piano and Fossa calderas (dashed lines enclosing divisions with the corresponding names) and the summit of the Fossa cone (continuous circle). The location map (right) shows the main road (grey) running the length of the island. Numbered locations are: 1. Il Faraglione. 2. Access to the Pietre Cotte lava flow. 3. Start of path for ascent to Fossa summit. 4. Summit of the Fossa cone. 5. Principal lava domes of the Lentia group. 6. Section of faulted pyroclastic units emplaced against Monte Saraceno. Geology from Ventura (1994).

units reflect both the northerly migration of the centre of activity and also variations in the common style of eruptive behaviour.

South Vulcano is believed to have been a simple stratovolcano about 5 km across at sea level and rising to as much as 800–1000 m above sea level. Its exposed lavas and scoria deposits date back to 100 000–120 000 years ago. Activity culminated with the collapse of a summit caldera 2–2.5 km across. The remaining southern rim of the caldera (Caldera del Piano: "caldera of the plain") marks the inland margin of South Vulcano, which appears today as a horseshoe-shape construct defining the southern limits of the island (Fig. 4.2).

For the next 50 000 years, eruptions from the collapsed floor filled the Piano caldera with lava flows and pyroclastic fall and surge deposits. About 24 000–15 000 years ago, activity became concentrated on the northwest side of the caldera, producing the distinctive Lentia units, dominated by pink lava domes, together with fall deposits. Further collapse occurred shortly afterwards (15 000–8000 years ago, starting on the eastern half of the island) to form the Fossa caldera, which lies northwest of and cuts into the products filling the Piano caldera (Fig. 4.2).

During the past 6000 years, activity within the Fossa caldera has been focused in the region where the present Fossa cone stands. Two kilometres across at its base and rising to 391 m above sea level, the last eruption from the cone in 1888–90 destroyed most of what is now the port area and current activity is confined to hot fumaroles degassing around the summit.

Fossa is the product of a series of eruptive cycles. At least seven cycles have been recognized (Table 4.2), characterized by an initial expulsion of breccia (mostly lithic material), followed by many pyroclastic surges, later pumice fall and, finally, the extrusion of a viscous lava flow. The sequence illustrates the explosive clearing of the upper conduit as rising magma vaporized meteoric water (basal breccia) and an evolution in eruptive style from hydromagmatic explosivity (dilute pyroclastic surges) through magmatic explosivity (pumice deposits) to lava effusion. The named cycles in Table 4.2 can account for only some 200–300 million m^3 of material, about one-third of the total volume of the cone, and so represent a partial history of Fossa's growth.

Erosional features in the upper pyroclastic levels of each sequence show that the cycles were interrupted by significant periods of repose.

Table 4.2 Eruption cycles from the Fossa volcano.

Cycle	Age of final lava flow	Erupted volume (10^6 m^3)
Punte Nere (at least three intervening cycles)	c. 6000 years ago	Tephra 195, lava 3
Palizzi	c. 1600 years ago	Tephra 5, lava 0.6
Commenda	AD 785	Tephra 25, lava 2.6
Pietre Cotte	1739	Tephra 25, lava 2.4

(Note that the Fossa and Vesuvius cones are similar in size and that the latter was constructed within only a few hundred years; see Ch. 2). Erosional features and palaeosoils are absent between deposits within each cycle, suggesting that any pauses during a cycle were short lived, although hiatuses lasting for as much as decades may have occurred. Fossa's most recent lava flow (the Pietre Cotte flow) was effused in 1739. After a pause of 32 years, intermittent activity recommenced at the volcano to end only after the 1888–90 outburst. It is still unclear whether the 1771–1890 activity marks an idiosyncratic end to the previous cycle or is the prelude to a new cycle.

The youngest construct is Vulcanello ("little volcano"), which, emerging above sea level in 183 BC, joined with Vulcano in 1550 to form the island's northern peninsula. Consisting of three pyroclastic cones and associated lava flows, Vulcanello last erupted in the sixteenth century and, until 1878, was also the site of intense fumarolic activity.

Vulcano: the 1888–90 eruption

Despite sporadic activity since the eighteenth century, the 1888–90 eruption remains the only event for which consistent observations are available. It is also the type-eruption used by Giuseppe Mercalli nearly a century ago to define Vulcanian behaviour. Although the eruption produced no casualties, it caused extensive damage in the area of the main port and led to the decline of sulphur quarrying on the island.

Opening with throat-clearing blasts on 3 August 1888, the eruption soon settled into the rhythm that was to characterize its behaviour for the following one-and-a-half years: powerful detonations (that broke windows even in Lipari) every few minutes, accompanied by the ejection of breadcrust bombs (see below), as well as ash-laden plumes that rose to heights of 2–3 km. These features, together with sudden changes from sustained explosivity to extended intervals of complete calm and vice versa, have become diagnostic for the Vulcanian class of eruption. Most striking is the force of the ballistic ejections, which hurled bombs, metres in diameter, more than a kilometre distant, where they crashed around the main port, devastating the quarry and surrounding buildings. Some people taking refuge in caves excavated by quarrying found their exits blocked by the giant bombs and had to be dug out.

Much of the ash accumulated around the upper flanks of the Fossa was subsequently washed away in the form of rainfall-triggered lahars, leaving bare the upper northern slopes of the cone. The occasional appearance offshore of small banks of pumice, together with the rupture of submarine cables between Vulcano and Lipari, indicate that magma was also escaping from undersea vents. Indeed, submarine activity may have continued for at least 18 months after the Fossa had stopped erupting on 22 March 1890, having expelled a modest 6 million m^3 of ash and bombs.

Vulcano: the volcano today

During the past century, fumarole temperatures at the Fossa rim have fluctuated from a typical 100–200°C to as much as 700°C. Peaks in high temperature occurred in 1916–24 and 1985–95. Both episodes were accompanied by heightened volcanic seismicity and the second, in particular, caused serious speculation about an imminent eruption. Indeed, ferry companies serving Vulcano from Milazzo in Sicily were alerted to emergency evacuation procedures.

Seismic clustering suggests that a magma body may lie some 5 km beneath the Fossa volcano. The area is under continuous surveillance, and seismic and ground-deformation data are transmitted to the new operations centre on the island and, via Lipari, to monitoring centres in Sicily. Situated in the Porto Ponente, the Vulcano Operations Centre (named after the Italian geochemist Marcello Carapezza) is open to the public; a small projection theatre has also been opened in the main port (Porto di Levante) where visitors can watch a video describing the volcanological history of the island.

Excursions to Vulcano

Vulcano can be visited at all times of the year, although sea conditions may be rough during the winter months. The peak season (July and August) is best avoided if possible, since during this period the 250 regular inhabitants of the port area are joined by a daily population of some 10 000 visitors.

Getting to Vulcano

Vulcano can be reached by car ferry (traghetto) and hydrofoil (aliscafo) from Milazzo in Sicily and also from Naples (Ch. 2). Milazzo is well connected by rail and road to Messina (and from there to the Italian Peninsula), as well as to Catania and Palermo and their respective airports.

Vulcano: the Fossa cone

The northern ascent of the Fossa cone passes through a range of surge deposits and erosional features. The first stop before ascent gives a rare opportunity to see the detailed structure of a rhyolitic (virtually obsidian)

Excursion time   By foot 3–4 hours ; allow 2–3 hours for the Fossa cone.

Maps   Istituto Geografico Militare topographic maps: 1:25 000 Sheet 244/iii/ se/Isola di Vulcano; 1:50 000 Sheet 581–586.1:25 000 topographic maps of all the main islands on a single tourist sheet, published by Trimboli.

Difficulty   Moderate for the Fossa cone, which can be climbed comfortably in 60 minutes. The path has been made from compacted ash; stout shoes are advisable. The best time to ascend is early morning, to avoid the scorching afternoon heat. **Whatever the time, ensure that plenty of water is carried**. Be prepared for short-lived but strong gusts of wind. Remember that the active fumaroles are likely to be degassing at temperatures exceeding 150°C and that the terrain close to the outlets can be friable. Access to the summit may be prohibited during periods of intense fumarolic activity.

lava flow. The summit provides spectacular views north across Vulcanello to Lipari and south over the Piano caldera. On clear days, gases can be seen rising over distant Stromboli.

From the main port (Porto di Levante), follow the road immediately on the left through the town. Continue beyond the shops (the end is marked by a post office on the left, after which the road is flanked by residential villas) for about 200 m, where a track opens to the left next to a small generating station. Follow the path where it turns sharply to the right and runs parallel with the road. About 30 m along, the path is flanked by slightly raised banks of monotonous mud-coloured deposits (a fine ochre-yellow matrix with millimetric black and grey lithics) containing lenses of angular fragments centimetres to decimetres across, dominated by black obsidian. About 70 cm thick, the bank nearer the road reveals stratified levels near its base. The layers range from a few millimetres to centimetres thick and consist of material similar to the matrix of the overlying, unstratified level. The whole bank is composed of flow deposits: the stratified basal units represent a series of pyroclastic surges, while the upper unstratified level reflects emplacement of a flow containing a greater proportion of solid particles. The lithic lenses indicate the disruption of chilled lava, probably originating from the conduit walls; the angularity of the lithics further suggests rapid shattering, possibly triggered by hydromagmatic explosions as heat from new magma flashed groundwater into steam.

It is worth pausing to inspect the textures shown by the larger lithic fragments. These range from black glass, through banded glass (black glass containing grey layers millimetres thick) to grey-transparent samples of highly vesicular glass. At first glance, it might be supposed that the range represents the transition from glass to pumice among pyroclastic material. Examination at the next location will dispel this interpretation.

Looking across to the Fossa cone, the most prominent feature is the toe of a thick, stubby lava flow that has just managed to descend the flanks of the volcano. This is the Pietre Cotte ("cooked stones") lava flow, effused from the lower northern rim of the Fossa crater in 1739. Negotiating a path through the bushes, it is possible to approach the base of the lava flow, which is here tens of metres thick. The obvious first impression is how the lower levels of the flow have broken along curved surfaces, giving the exterior a flaky appearance. **Be careful when close to the flow; the small scree slopes below the base bear witness to the instability of the exposed section.**

The Pietre Cotte flow (2.4 million m^3) was produced during the final stage of Fossa's 1731–9 eruption. It is a glassy K-rhyolitic lava, whose very high viscosity is responsible for the stubby form of the flow, even though it rests on the steep flanks of the Fossa cone. Inspection of debris from the flow reveals the same range in textures as seen among the lithics from the material flanking the path at the previous stop. (The lithics are clearly not derived from the overlying – and hence younger – Pietre Cotte flow, but from an earlier, similar type of lava flow.) The rounded contortions of the thin grey bands within the glass can thus be related to flowage as the lava descended the cone. Compared to the black glass, the bands appear grey because they contain a high proportion of tiny bubbles (best seen with a hand lens), an association to be expected given the vesicular nature of the transparent-grey samples (some of which, incidentally, are light enough to float on water).

Since glass is an inherently unstable form of solid, local viscous heating during lava advance must have been just sufficient to trigger the exsolution of dissolved volatiles as bubbles. Accordingly, the vesicular bands trace planes of concentrated deformation within the lava flow. Zones of extreme vesiculation are common near the base of the lava, since this is the part in which deformation is concentrated as a flow overcomes frictional resistance from the ground. An important lesson here is that apparently pumiceous samples can in fact be formed in lava flows with evolved compositions. Hence, although the lithic fragments from the previous stop might have been interpreted as the products from a pumice-forming eruption, in fact they were produced by the slow effusion of (in this case) obsidian lava flows.

From the Pietre Cotte flow, return to the road and continue in the direction away from the port. After about 600–700 m, a brown sign indicates a path (near a giant cypress pine tree) to the left for the ascent of the Fossa cone. To begin with, the path follows long gentle zig-zags (which can be short-circuited by the energetic) along a path of compacted scoria. About three quarters of the way up, the terrain changes to pink, compacted and very fine ash (for the most part dust and ash from old material thrown out during the 1888–90 eruption), heavily gullied and irritating to negotiate.

On the way, look out for sporadic outcrops of stratified, or even cross bedded, pyroclastic surge units.

At the summit, the most evident feature is the nested crater, the innermost of which was the source of the last eruption in 1888–90 (Fig. 4.3). Normally, some wag has left a message in stones on the crater floor. (Frankly, descent into the crater is time-consuming and pretty pointless – unless, of course, you want leave a message to annoy later visitors.) The second obvious feature is the series of fumaroles degassing from the rim, leaving behind a trail of yellow and white sulphur compounds. Remembering that the gases are likely to be emerging at temperatures exceeding 150 °C, it is nevertheless worth investigating the deposition of fine laths of raw yellow sulphur around the points of gas emission. (The gases may smell, but are usually bearable and, should problems arise, they can be fixed by a handkerchief to the nose and mouth. In extreme cases, noxious gases can be absorbed by material soaked in urine, but that option is voluntary.)

Allow 30–40 minutes for a tour of the crater rim. On the northern side, by far the greatest attraction is the view across Vulcanello to Lipari and, weather permitting, to Salina (with two peaks) beyond and to the left. On exceptionally clear days, gases can be seen rising from Stromboli (beyond Lipari to the right). The highest point of the Fossa cone, on the south, also offers splendid views of the Piano caldera and younger, infilling products, as well as, to the northwest, the upstanding rim of the island defined by the Lentia rhyolite domes. Above all, it is easy to appreciate that the whole of the tourist complex around the main port is vulnerable to the effects of the next eruption.

After traversing the summit rim, the volcano can be descended following the path of ascent. Before leaving the summit area, look at the nature of the deposits across the surface. These consist of scoria and finer material, mixed with many bombs reaching metres across. Although the best samples have already been taken away, the bombs show a characteristic "bread crust" exterior, that is a segmented form as found on bread taken directly from the oven (commercial sliced bread does not count). This pattern results from the cooling of viscous magmatic ejecta while hurtling through the air before impact with the ground. The texture is attributable in part to rapid thermal contraction of the bomb's exterior and also to the escape of gases temporarily trapped beneath the chilling surface. It is diagnostic of viscous magma and it contrasts with the smoother surfaces and aerodynamic shapes commonly assumed by bombs of more fluid magma, as can be found on Stromboli and Etna.

Figure 4.3 The Fossa crater, looking north towards Lipari in the background.

Vulcano: from the sea to the interior

A single road leads from the main port to the southern end of the island. The northern 4 km of the road from the main port to Monte Saraceno provides good views of large-scale features from the Fossa cone and Lentia K-rhyolite domes to the rim of the Piano caldera.

Excursion time   Three to four hours on foot.

Maps   As for Fossa cone itinerary.

Difficulty   Easy. All walking is along roads.

Leave the main port along the road to Gelso and the south of the island. Continue beyond the track leading to the Fossa cone (previous excursion). To the right, the high rim of the island is delineated by the partially obscured pink K-rhyolite lava domes of the Lentia group (compare the colour with that of the K-rhyolitic Pietre Cotte flow). The most southerly large outcrop, for which major collapse has dissected the dome, reveals a pattern of almost radial curved joints (Fig. 4.4). The jointed columns formed by inward cooling of the dome. Radial contraction has forced the lava to break along planes extending from the surface to the dome centre (strictly, the jointing is quasi-radial, in that the joints meet along a planar horizon and not a point at the dome's centre). About a kilometre beyond the route to the Fossa cone, at least two tracks on the right lead up to the domes which, at close hand, show both planar and scoriaceous surfaces and a broadly concentric arrangement of vesicles (allow an extra 60–90 minutes to visit the domes).

Figure 4.4   Rhyolite lava dome of the Lentia Group. Collapse has revealed an almost radial columnar jointing across the dome interior.

The left of the road provides a good panorama of the Fossa cone within the Fossa caldera. The south flank of the cone is vegetated to the summit, but wind erosion has stripped plant cover from the upper third of the northern flank, exposing the pink 1888–90 ashes as a step-like bald patch with a well developed dendritic pattern of gullies cut by the surface runoff of rainwater (Fig. 4.5). Ravens may be seen gliding over the southern plain between the cone and caldera walls, for which the district is known as the Timpone di Corvo. A bulge in the southwest midflanks of the cone marks the rhyolitic Commenda lava flow (2.6 million m^3), possibly erupted in AD 785. Unaccountably, part of the timpone is used as a rubbish tip.

Beyond the path leading to the Fossa cone, the road continues to ascend gently for about 3–3.5 km before reaching a tight right-hand bend on the inland flank of Monte Saraceno which, at 481 m above sea level, is the second highest point on the island (the highest point is Monte Aria at the southeast end of Vulcano). Here the road follows the rim of the Fossa caldera. Just before the bend, the view on the left shows the lavas and pyroclastic units that filled the Piano caldera before the collapse of the Fossa caldera. Solar panels around the base of the Fossa cone pinpoint the locations of automatic monitoring stations.

The right-hand side of the road reveals a faulted sequence of pyroclastic units emplaced against the slopes of Monte Saraceno when this peak formed part of the exposed Piano caldera wall. Some 25–30 m of outcrop can be viewed from the road (Fig. 4.6). It consists of about 4–6 m of pink and yellow material, sandwiched between thicker units of grey and grey–pink fragmented rock. The main face of the pink and yellow units shows

Figure 4.5   The Fossa cone, looking east. Wind and precipitation have preferentially stripped away the surface layers from the cone's upper northern flanks.

Figure 4.6 Faulted sequence of pyroclastic units emplaced against the flanks of Monte Saraceno. Faulting (downthrow to the left) is seen most clearly in the pink (dark) and yellow (light) pyroclastic-surge deposits. The irregular surfaces to the top and bottom of the photograph define scoriaceous units emplaced either by intense lava fountains or as a pyroclastic flow. The pink pyroclastic level on the downthrow side of the fault is about 3 m thick.

four layers, each about 1 m thick, picked out by changes in colour and also by local concentrations of angular black lithics, decimetres across, along the base of some layers. Following these layers to the sharp bend in the road (taking account of the downward slipping of the rock face), it is possible to glimpse sections with a different orientation that have picked out finer layers centimetres thick. The fine layering and basal accumulation of lithics suggests the deposits were produced by pyroclastic surges or by dilute pyroclastic flows, although the distinction here is largely semantic.

Of the grey units, only the lower is accessible, at the bend in the road. It consists of poorly sorted, moderately vesicular and irregular fragments decimetres across, among which are a few bands of apparently continuous trachybasaltic lava, about 1 m thick and tens of metres long. The fragments contain a fine network of tiny cracks and, when crumbled in the hand, they release euhedral, greenish-black pyroxene phenocrysts (millimetres long) from the disintegrated matrix (note the analogy with the final products from Vesuvius' 1944 eruption; see Ch. 2).

The grey unit resembles a section along the margin of a strongly auto-brecciated aa lava flow (compare with road sections through aa flows on Etna and Vesuvius). Indeed, combined with the apparent dip into the Piano caldera, this material was originally interpreted as a lava flow from Monte Saraceno. However, there are no vents on Monte Saraceno, which owes its conical form to the mode of collapse during formation of the Piano caldera. Further inspection of the outcrop shows, in addition to the apparently continuous lava flows, the presence of large bombs, metres in diameter, of massive lava. An alternative interpretation, therefore, is that the unit consists of a mass of bombs and scoria, sometimes welded into apparent lava units, that was erupted from a centre within the Piano caldera and emplaced against the exposed caldera wall (so that layers dip into the caldera). The precise eruptive mechanism remains uncertain between a type of flow or vigorous lava fountaining. Whatever the mechanism, the picture that emerges is of intense explosive activity from vents within the Piano caldera, producing a combination of dilute pyroclastic flows and lava fountains or dense currents of trachybasaltic fragments.

Monte Saraceno is the last stop of the excursion. From here, it is an easy descent along the road back to the port. Continuing south, the double-bend winding up hill passes through a floral extravaganza before reaching the southern rim of the Piano caldera, where the ground levels out and gives way to extensive agricultural land.

Vulcano: other sites to visit

Il Faraglione
Immediately behind the main port, a newly paved brick path leads across Il Faraglione ("the stack"), the remnants of a multicoloured pyroclastic cone (whose hues are the result of fumarolic alteration), towards the famous mudpool of Vulcano. In the nineteenth century (until the 1888–90 eruption) this area was quarried, mostly with convict labour, for its sulphur and potassium compounds. Shortly before the most recent eruption, the quarry had been taken over by James Stevenson, a Scot who used the raw material in his chemical factory in Glasgow. The mudpool is a disused quarry pit, only weakly recharged by natural circulation (i.e. the same water has already been used by hundreds of others beforehand). Although there may be the vicarious thrill of bathing in slimy mud 5 km above a magma chamber, there are social effects to be considered. The pool stinks of decaying paté and, once the mud has infiltrated the skin, it is an aroma that remains for at least a couple of days. Those who go into the pool should be prepared to smell like yesterday's leftovers from someone else's dinner – the part that even a dog has refused. Rather more salutary are the fumaroles heating the sea along the beach just north of the mudpool.

Vulcanello
Between the main port and the Porto di Ponente, a road leads northeast to Vulcanello. The volcano consists of three overlapping pyroclastic cones,

the youngest of which was quarried for sulphur and potassium compounds. A trachytic lava can be seen at Punta del Roveto, near a camping
site on the northeast coast.

Around the island
Many opportunities are available to travel around Vulcano in a small boat
(about 2 hours). The trip may be combined with a visit to Lipari.

Vulcano: Further reading

Websites
For the volcanological history of Vulcano and the Aeolian Islands:
  http://www.geo.mtu.edu/~boris/vulcano/
For the Istituto Internazionale di Vulcanologia in Catania:
  http://www.iiv.ct.cnr.it/
For the Poseidon surveillance network of Sicilian volcanoes:
  http://www.poseidon.nti.it/

Literature
De Astis, G., L. La Volpe, A. Peccerillo, L. Civetta 1997. Volcanological and petro-
  logical evolution of Vulcano (Aeolian arc, southern Tyrrhenian Sea). *Journal of
  Geophysical Research* **102**, 8021–8050.
Frazzetta, G., P. Y. Gillot, L. La Volpe, M. F. Sheridan 1984. Volcanic hazards at Fossa
  of Vulcano: data from the last 6000 years. *Bulletin Volcanologique* **47**, 105–124.
Keller, J. 1980. The Island of Vulcano. *Rendiconti della Società Italiana per Mineralogia
  e Petrologia* **36**, 369–414.
Montalto, A. 1996. Signs of potential renewal of eruptive activity at La Fossa
  (Vulcano, Aeolian Islands). *Bulletin of Volcanology* **57**, 483–92.
Sheridan, M. F., G. Frazzetta, L. La Volpe 1987. Eruptive histories of Lipari and
  Vulcano during the past 22000 years. In *The emplacement of silicic domes and lava
  flows*, J. H. Fink (ed.), 29–33. Special Paper 272, Geological Society of Amercia,
  Washington DC.
Ventura, G. 1994. Tectonics, structural evolution and caldera formation on Vulcano
  Island (Aeolian archipelago, southern Tyrrhennian Sea). *Journal of Volcanology and
  Geothermal Research* **60**, 207–224.

Stromboli

About 50 km north of Vulcano, Stromboli is the most northerly and most
active of the Aeolian archipelago. For at least 2000 years, vents at the summit have been ejecting magma with almost clockwork precision, offering
to sailors a natural beacon by night and for which Stromboli has become
known as the Lighthouse of the Mediterranean.

Although reaching 924 m above sea level, the island measures only

4 km by 3 km, endowing it with an almost pyramidal shape whose slopes lie typically between 25° and 30° (Figs 4.7, 4.8). For the past 100 000 years, the island has undergone repeated episodes of construction, summit collapse and major flank failure. The scar of the most recent giant landslide, from 5000 years ago (an age similar to that of Etna's Valle del Bove), is still clearly visible along the northwest coast. With its headscarp cutting the summit, the landslide has left a U-shape gouge 1–1.5 km across that extends below sea level for another 700 m. Captured by this natural depression, lava flows from the summit frequently illuminate the valley an incandescent red, from which the scar has earned the name Sciara del Fuoco ("descent of fire").

The island itself is but the tip of a volcano. Ninety-eight per cent of the volcanic edifice lies under water, extending to depths of 1.5–2 km. (To envisage this, imagine Etna flooded to the top of the Valle del Bove.) As a result, much of the observed activity may well represent a fraction of the events occurring over the whole volcano. Indeed, the oldest exposures (2.3 million years old) of Stromboli's activity come not from the main island but from Strombolicchio, a tiny volcanic neck about 2 km to the northeast. The neck (which can be visited by boat) consists of the remnants of an earlier feeding conduit filled with solidified magma (the surrounding scoria of the edifice has since been eroded by the sea). However, it is not an independent construct, but is connected to Stromboli's submarine flanks.

Following the exhaustion of Strombolicchio, growth of the main island can be divided into four main stages (Table 4.3): Palaeostromboli, Vancori, Neostromboli and Recent Stromboli. The stages are distinguished in part by variations in the chemistry of erupted magma, and in part by closing events that involved major collapse at the summit or down the volcano's flanks. In all stages, eruptions appear to have been concentrated in the vicinity of the present summit (although there is some evidence for a modest northwest migration of the vents). Rare examples of near-coastal eruptions include the low west-flank 1768 eruption (Fig. 4.8), as well as the prehistoric vents of Neostromboli at the Timpone del Fuoco (near Ginostra) and Nel Cannestrà (near Ficogrande).

Since the formation of the Sciara del Fuoco, Stromboli's subaerial

Table 4.3 Principal stages in the subaerial growth of Stromboli.

Stage	Age (thousand years ago)
Strombolicchio	230
Palaeostromboli	35 –100
Vancori	13 –26
Neostromboli	5.5 –13
Recent Stromboli	< 5.5

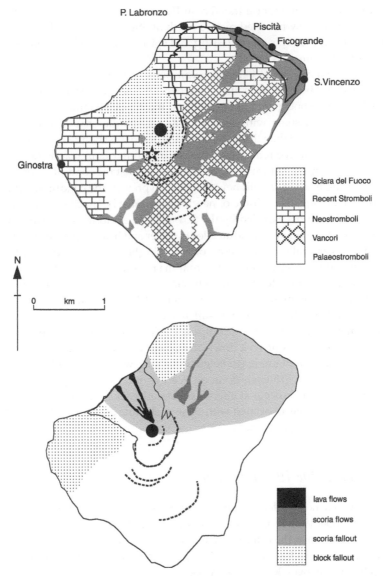

Figure 4.7 The four main volcanological divisions of Stromboli. Dashed lines show collapse rims. The zone of active craters (black circle) lies just northwest of the viewpoint Pizzo Sopra la Fossa (between the innermost collapse rims) and north of I Vancori, the volcano's summit (star). The coastal route (black) reaches Pizzo Sopra la Fossa after skirting alongside the Sciara del Fuoco. The lower map shows the distribution of material expelled during the eruption on 11 September 1930. Geology from Hornig-Kjarsgaard et al. (1993). Map of the 1930 deposits from Rittmann (1931).

activity has become concentrated near Pizzo Sopra la Fossa ("peak above the crater"), a few hundred metres north of Stromboli's peak, I Vancori. Currently, three vents are degassing and episodically expelling bombs to heights of as much as 200 m.

The regular interval between explosions is attributable to the time required for bubbles to collect just beneath the magma surface and, coalesced into one giant bubble, to burst the chilled magmatic skin. The ejected bombs, some with an aerodynamic spindle shape denoting a magma more fluid than that producing the breadcrust bombs on Vulcano, are thus the ruptured remnants of the wall of an enormous magmatic bubble. The persistent nature of the activity means that the conduit is continually being recharged with new magma. To prevent the conduit magma from solidifying, the volume of new magma must be much greater than the small amounts expelled by the Strombolian explosions. Apparently, convection in the upper conduit allows cooler and heavier magma to sink into a larger shallow-level reservoir (within the submarine part of the volcanic edifice) as new and hotter magma rises to take its place.

Such predictable steady-state behaviour is occasionally interrupted by hiatuses of a few months (and, very rarely, of 1–2 years) without surface activity (although magma may still reside within the volcano) or by a dramatic increase in magmatic output. The larger eruptions may be dominated by effusions of lava down the Sciara del Fuoco (Fig. 4.8), or by vigorous lava fountaining (to heights of 1 km), with tephra fall and also pyroclastic flows. Effusions occur on average once every 4–5 years; major explosive outbursts occur at intervals of decades. The explosive events, in particular, may be associated with seismic shaking and landsliding, both of which may trigger tsunamis.

Stromboli: the 1930 eruption

Stromboli's most violent historic eruption occurred without warning on 11 September 1930. It lasted for less than a day, but left behind six dead and 24 injured among the ruins of the Ginostra and Stromboli villages; it also catalysed an exodus from the island, whose population since 1930 has declined from over 3000 to a mere 350.

Previous activity that year had been nothing out of the ordinary: persistent mild explosions and, in February, modest lava emission. The onset of unusual behaviour was first noticed at 08.10 h when ten minutes of ash emission left a veil of deposits over the southwest of the island. A hundred minutes later (09.52 h), the edifice was rocked by two giant explosions that hurled lithic blocks, metres in diameter, as far as Ginostra and Punta

Figure 4.8 Stromboli erupting in 1768. Lava is descending from the summit craters down the Sciara del Fuoco (left). Gases from flank vents can be seen low on the right. Painting from William Hamilton's "Campi Phlegraei" (1776).

Labronza, both 2 km away. At the same time the sea pulled back from the coastline, to return shortly as a train of tsunamis 2–2.5 m high; even around Lipari boats were carried ashore.

As the shaking died away, fountains of gas-rich magma began playing from the vents, piling-up thick and unstable accumulations of scoria to beyond the rim of the Sciara del Fuoco. Within 45 minutes, lapilli were coating Stromboli village to depths of 10–12 cm, while collapses of the unstable accumulation at the summit sent two flows of incandescent scoria streaming down the Vallonazzo Valley and into the sea near Piscità. Explosive activity had all but finished by 10.40 h, the eruption giving way to the outpouring of lava flows down the Sciara del Fuoco until after dusk.

The consequences of the eruption were devastating. Four fishermen were killed offshore by the scoria avalanches entering the sea; one person died beneath blocks falling over Stromboli village, and another was claimed by the tsunamis. Roofs in Ginostra and Stromboli were crushed by blocks and piles of tephra, which also set vegetation alight.

Eruptions such as the 1930 event are expected every few decades. Of greatest concern is the apparent lack of precursory signals. The abrupt onset of the 1930 sequence may have been triggered by the arrival of new magma that caused the edifice to swell (and the sea to withdraw) and to open a lateral undersea fissure. Rapid drainage of magma from the summit into the new fissure left the upper conduits unsupported. Parts of the conduit began to collapse, allowing the infiltration of water, which, flashing to steam, provoked the two giant explosions at 09.52 h. With the upper conduits again clear, new gas-rich magma reached the summit to feed the lava fountains. Finally, having exhausted the gas-rich component, the eruption waned with the effusion of lava.

Stromboli: the volcano today

A key challenge is to identify precursors to the volcano's large explosive events. The interpretation of the 11 September 1930 eruptive sequence suggests that warning signals may be forthcoming from seismic and deformation data as new magma enters the undersea portion of the volcanic edifice. As a result, Stromboli is today under continuous geophysical surveillance, information being sent to the monitoring station in San Vincenzo (Stromboli village) and also to centres in Sicily.

Stromboli: excursions

Travel details as for Vulcano.

Ascent of Stromboli

The summit craters can be reached by two routes from Stromboli village. One route cuts across the centre of the island. The other follows the coast to Punta Labronzo and ascends from there. The simpler Punta Labronzo route is followed here (Fig. 4.7). Further details, including information on other routes, can be obtained from the Azienda Soggiorno Isole Eolie (Aeolian Island Tourist Board) on 090-988.0095.

Excursion time Allow a full day (the ascent and descent together take about five hours at least). Check with local guides about the safety of overnight stops near the summit crater; measures are under way to prevent such stops without a guide, although in practice there may be some laxity in enforcement. **Remember that Stromboli is NOT a "safe" volcano; unwary visitors have been killed and injured in the summit area.**

Maps Istituto Geografico Militare topographic maps: 1:25 000 Sheet 244/I/SE; 1:50 000 sheet 577 bis.1:25 000 topographic maps of all the main islands on a single tourist sheet, published by Trimboli.

Difficulty Moderate to difficult. The upper half of the ascent is very steep on poorly compacted terrain and without shade. Those with respiratory or cardiac difficulties should seek medical advice about ascending to altitudes of 800–900 m above sea level. **Boots, food and water are essential.** Windproof clothes are advisable at the summit.

From the port in San Vincenzo (services from Sicily and the other islands), pass through Stromboli village to Piscità on the north coast (services from Naples may stop here). Continue beyond the village along the well marked path that follows the coast for about 1.5 km before winding inland (at Punta Labronzo) through fields of bamboo and scrubland towards the eastern rim of the Sciara del Fuoco (Fig. 4.7). The path of poorly compacted ash and scoria runs alongside the landslide scar for almost 1 km, providing excellent views of the lava flows covering the floor of the Sciara. Look out also for exposed dykes and sections showing graded tephra deposits.

The path continues about 500 m beyond the headscarp of the Sciara until it reaches Pizzo Sopra la Fossa (PSF), from where the Strombolian activity can best be observed. The active vents are set inside three craters in a terrace below and to the north of PSF. **On no account should they be approached, even if the vents appear to be inactive: the chances are they won't be so for more than 20–30 minutes**. Cavalier behaviour puts not only yourself at risk but also those who may have to collect your remains if events go against you. It should also be remembered that, even though a few hundred metres away, the observing point at PSF is well within the range of bombs from stronger (albeit infrequent) explosions.

The return to Stromboli village is by the same path as for the ascent. To catch the red glare of explosions, it is popular to ascend the volcano in late afternoon and to remain at the summit after sunset. In this case, travelling in a group is advisable and extra time should be allowed for descent in the dark (take torches). Alternatively, arrange an evening at the Osservatorio pizzeria at Punta Labronzo, which boasts a good view of the Sciara del Fuoco. If returning by day, a refreshing stop can be had at the black-sand beach of Ficogrande (remember that the volcanic sand can become scalding hot in the afternoon sun).

Stromboli: other sites to visit

Small boats are invariably available for tours around the island. These offer excellent views of the Sciara del Fuoco, especially at night. Once in a boat, it is also worth visiting Strombolicchio (a rusty staircase leads to the top) and the isolated village of Ginostra on Stromboli's southwest coast.

Stromboli: further reading

Websites
For the volcanological history of Stromboli and the Aeolian Islands:
 http://www.geo.mtu.edu/~boris/stromboli/
 http://vulcan.fis.uniroma3.it/gnv/
For the latest news on Stromboli, including "livecam" video of the summit craters:
 http://educeth.ethz.ch/stromboli/index-i.html (Stromboli On-Line)
For the Istituto Internazionale di Vulcanologia in Catania:
 www.iiv.ct.cnr.it
For the Poseidon surveillance network of Sicilian volcanoes:
 http://www.poseidon.nti.it

Literature
Anderson, T. 1905. On certain recent changes in the crater of Stromboli. *Geographical Journal* **25**, 123–38.
Hornig-Kjarsgaard, I., J. Keller, U. Koberski, E. Stadlbauer, L. Francalanci, R. Lenhart 1993. Geology, stratigraphy and volcanological evolution of the island of Stromboli, Aeolian arc, Italy. In Manetti & Keller (1993: 21–68).
Manetti, P. & J. Keller (eds) 1993. *The island of Stromboli: volcanic history and magmatic evolution*. Special issue of *Acta Vulcanologia* **3** [dedicated volume].
Rittmann, A. 1931. Der Ausbruch des Stromboli am 11 Sept 1930. *Zeitschrift für Vulkanologie* **14**, 47–77.

Lipari

The largest of the Aeolian Islands (37.6 km²), Lipari also boasts the largest population (8515) of the archipelago. It has been a commercial centre since Neolithic times, initially exporting obsidian across the southern Mediterranean and, latterly, trading its extensive pumice deposits.

At the junction between the tail and the main body of the squat Y-shape arrangement of the subaerial Aeolian islands (Fig. 4.1), Lipari has erupted magmas from all the principal Aeolian trends: basalt to rhyolite (and its potassic equivalents) and K-basalt to K-trachyte. With a foundation that reaches a kilometre below sea level, Lipari's earliest surviving subaerial products are about 220 000–160 000 years old (Table 4.1). Dominated by lavas and domes (basaltic to andesitic), these products line the west coast of the island (Fig. 4.9). For the following 70 000 years (until about 90 000 years ago), eruptions built up what is now the main part of the island to the south and west. The erupted products range from lava flows and domes to thick sheets of pyroclastic surges, fed by magmas of the basalt–dacite (and potassic) trends. Prominent features this period include Monte Sant'Angelo (594 m above sea level), Monte Chirica (the highest point on the island at 602 m above sea level) and Quattropani (Fig. 4.9).

Between about 42 000 and 20 000 years ago, a new period of activity fed

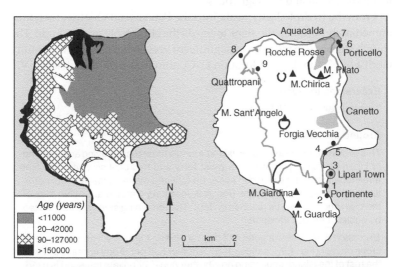

Figure 4.9 Sketch geological and location maps of Lipari. The geological map (left) shows the four main divisions based on the period of activity. Erupted materials within each division consist of lava and pyroclastic deposits. The location map (right) shows the circum-island main road (grey). Numbers refer to locations described in the text. The local peaks are shown as filled triangles. Also shown are the historical obsidian flows (pale grey) and collapse rims (black). Geology from Crisci et al. (1991).

the rhyolitic domes of Monte Guardia and Monte Giardina, on the south of the island. Some 10 000 years later, activity became focused to the north-east, creating the rhyolitic Monte Pilato ("bald mountain") complex (built up intermittently between about 11 000 BP and AD 729). The last episode of activity (during the interval AD 580–729) produced the giant pumice cone of Monte Pilato (476 m above sea level), whose summit crater is 1 km across and nearly 200 m deep. The end of the eruption was marked by the slow effusion of the Rocche Rosse ("red rocks" – the colour is attributable to surface weathering) obsidian lava flow, which today forms the island's northeastern promontory. At about the same time, obsidian lavas were also emerging from Forgia Vecchia, some 3 km farther south on the island's eastern coast (Fig. 4.9).

The clearest sign of volcanism today is the intense fumarolic activity off the west and east coast of the island. However, since historical and geo-logical studies suggest that Lipari tends to cluster periods of eruption at intervals of between several centuries and millennia, it is likely that erup-tive episodes will occur again. In the meantime, spectacular deposits of pumice and banded rhyolite can be found across the island, together with trachybasaltic scoria and lavas, including hyaloclastites. The Museo Eoliano in Lipari town (via del Castello; 090-988.0174/0594) also reminds visitors of the intimate dependence between volcanism and human activ-ity on the island and its neighbours.

From the hydrofoil jetty, walk into the Piazza Sant'Onofrio, then turn immediately south up the steps towards the church of San Guiseppe. The church steps are made from a magnificent grey welded tuff and certainly merit a brief stop (Fig. 4.9, location 1). Continue past the church along Via

Excursions to Lipari

Travel details as for Vulcano.

Around the island

This circular itinerary follows the island's single main road, leaving and finishing at Lipari town, the port-of-call for most ferry and hydrofoil services. The first three locations can be reached easily on foot and make a pleasant tour of the town itself. Two other locations (Fig. 4.9, locations 4 and 5) are within 1 km of the town and can also be reached on foot. The remaining stops are best visited by public or private transport.

Excursion time Allow a good half day for visiting Lipari town and nearby out-crops on foot (Fig. 4.9, locations 1–5). Another half or full day is needed to visit the rest of the island, depending on whether public or private transport is used.

Maps Istituto Geografico Militare topographic maps: 1:25 000 sheet 244/III/NE; 1:50 000 sheets 581–586 (Lipari and Salina). 1:25 000 topographic maps of all the main islands on a single tourist sheet, published by Trimboli.

Difficulty Easy.

Maddalena and down the hill past the Hotel Rocce Azurre to the small beach of Portinente. Cross the beach and climb the hill on the south side of the bay to the obvious outcrops of welded tuff breccia (location 2) exposed on the flank of the Punta San Guiseppe rhyolite dome (about 20 000 years old).

Return across the beach to Piazza Sant'Onofrio and walk northwards along Via Garibaldi. Take the steps on the left-hand side and climb to the cathedral and museum (location 3), both of which are built on another massive rhyolite dome (about 20 000 years old). The widely exposed rock is well-jointed and often flow foliated. If time permits, visit the museum for an overview of how Lipari's geology has dictated the economic growth of the Aeolian Islands.

Leaving the museum, continue north and descend to the ferry port (Approdo Traghetti). Take Via Amendola along the west side of Porto Sottomonastero. Large blocks of porphyritic andesite can be seen on the beach (location 4). Continue along the road, following the signs for Canetto and Porto Pignataro. About 100 m before the road tunnel, fresh outcrops of medium- to thin-bedded andesitic hydromagmatic rocks (100 000 years old) occur by the road and in cliffs beside a small path up to a house (location 5). Grey lapilli tuffs, in places cross bedded, and ash layers enriched in lithics are also evident. Near the tunnel (northwards), the tuff sequence is overlain by andesitic lavas with subspherical flow foliation.

The walk can be continued through the tunnel to Cannetto, from where it is possible to take a service bus to the quarries and beach at Porticello. However, it is probably simpler to hire either a car or scooter to complete the itinerary.

The coast road skirts Monte Pilato and provides excellent views of the pumice beds and brown-weathering obsidian flow for which Lipari is famous. At Porticello take the narrow road down on the right to the jetty (location 6), which provides excellent views into the quarries of bright white pumice, emplaced during the medieval activity of Monte Pilato (Fig. 4.10). Walk a short distance northwards along the beach to inspect obsidian outcrops, both on the beach and in the toe of the massive Rocche Rosse lava flow. Note the contact between pumice and rhyolite lava, as well as the several stages of devitrification within the obsidian itself. Returning to the road, additional excellent exposures of obsidian can be found in a small quarry opposite the turnoff (location 7).

The circum-island road continues through Aquacalda and climbs towards Quattropani. Before it leaves the coast there is a pullout on the right-hand side (location 8), which, on a clear day, affords a spectacular view of the Aeolian arc from Alicudi in the west to Stromboli to the northeast. The road continues across a low caldera to Quattropani,

Figure 4.10 Coastal pumice quarry at Porticello, just south of location 6 in Figure 4.9. The brilliant white pumice was deposited during the medieval activity of Monte Pilato.

characterized by abundant outcrops of massive, pumiceous pyroclastic deposits (about 8600 years old), some of the units containing dune-like bedforms (location 9). On the return to Lipari town, the road continues through Varesana and Pianoconte. About 1.5 km south of Pianoconte, wonderful views open out over the southern coast of Lipari towards Vulcano. Once back in Lipari town, the day's observations can be reflected upon over a glass of Malvasia di Lipari, a sumptuous wine of which the islanders are justly proud.

Lipari: other sites to visit

Around the island
Starting from Lipari town (also San Pietro on Panarea), boat daytrips are available for visiting Lipari, Panarea, Basilluzzo and Stromboli. Contact the Pignataro Shipping Service (090-981.1417, mobile 0368-675975).

Lipari: further reading

Websites
For an overview of Lipari's volcanology with images:
http://www.geo.mtu.edu/~boris/lipari/
For general tourist information:
http://www.tau.it/aapitme/isoleen.html

Literature
Crisci, G. M., R. De Rosa, G. Lanzafame, R. Mazzuoli, M. F. Sheridan, G. G. Zuffa
1981. Monte Guardia Sequence: a Late-Pleistocene eruptive cycle on Lipari (Italy).
Bulletin of Volcanology **44**, 241–55.
Crisci, G. M., R. De Rosa, S. Esperanca, R. Mazzuoli, M. Sonnino 1991. Temporal
evolution of a three component system: the island of Lipari (Aeolian Arc, southern
Italy). *Bulletin of Volcanology* **53**, 207–21.

Panarea

Panarea (Fig. 4.11) is the celebrity island of the Aeolian arc. For most of the
year, about 300 people peacefully share its $3.5 \, km^2$. However, bolstered by
the glitter of the rich and famous, the population swells to over 2000 during the summer. For visits that are meant to be volcanological, it is therefore prudent to avoid the period from late July to early September –
especially as prices also increase as much as the population does.

Rising to 421 m above sea level (at Punta del Corvo), Panarea is part of
the summit rim of a volcano that extends to depths of 1200–1700 m. It
appears to be connected under water with Stromboli, 45 km northeast, a
neighbour with which the volcanic products from Panarea share a similar
chemical diversity. Unlike Stromboli, however, Panarea is dominated by
lava domes, with only a modest presence from lava flows and pyroclastic
deposits.

The earliest exposed products are the K-dacitic lava flows at Punta
Muzza, on the island's west coast, effused about 150000 years ago. Subsequent effusions (K-andesite to K-dacite) had established the bulk of the
present island within 25000 years, after which local activity appears to
have ceased for some 65000 years. Activity resumed between 60000 and
13000 years ago, during which time basaltic and basaltic–andesitic scoria
and lapilli were erupted from centres in the east near Punta Falcone and
Punta Torrione. Since then, the island has lain dormant.

The hiatus in eruptions between 125000 and 60000 years ago may well
be more apparent than real, as regards activity from the whole volcano
beneath Panarea. The island itself is but the exposed western fragment of
a summit complex, about 11×8 km across and submerged to depths of 100–
150 m as a result of subsidence (demonstrated by the drowned Roman settlements that can be seen off shore). Tiny upstanding portions of this complex form the small collection of islets dotted eastwards (NE–SE) of
Panarea. Of these islets, the largest is Basiluzzo, part of a rhyolite dome
erupted at some time during Panarea's most recent phase of activity
(13000–60000 years ago). Thus, while basalts and basaltic andesites were
being erupted from Panarea, rhyolite magma was emerging at Basiluzzo,

Figure 4.11 Sketch location map of Panarea, showing the island peak (filled triangle), the main road (grey) and the locations (numbers) described in the text. The bulk of the island consists of lava domes and flows.

just 2.5 km to the northeast. Evidently, the submerged volcano has enjoyed a complex eruptive history, of which the rocks from Panarea and Basiluzzo tell only a part of the story.

East-coast itinerary

The itinerary follows the main island road from Punta Milazzese to Calcara. Hydrofoil services dock half way along the route at the jetty in San Pietro.

From the jetty, walk south, beyond the Hotel Lisca Blanca, onto the small headland of Punta Peppermaria, where excellent outcrops can be found of the rhyodacite lavas erupted about 130 000 years ago (location 1).

Excursions to Panarea   Travel details as for Vulcano.

Excursion time   Allow a day on foot.

Maps   Istituto Geografico Militare topographic maps: 1:25 000 sheet 244/I/NO; 1:50 000 sheet 577 bis. Gruppo Nazionae per la Vulcanologia, 1:10 000 geological map of Panarea and Basiluzzo (linked with Calanchi et al. 1999). 1:25 000 topographic maps of all the main islands on a single tourist sheet, published by Trimboli.

Return to the hotel, turn left and walk up the hill to join the circum-island road. Turn left (south) and follow the signs for the Bronze-age Villaggio Preistorico. A half-hour saunter through the village, which is a delight to explore, and across the Caletta dei Zimmari beach leads (after a final stiff but short climb) to the dramatic headland at Punta Milazzese (location 2). The cliffs have been etched out of beautifully jointed andesite lava flows cut by dykes. (As described below, the cliffs are best seen from a boat.)

Return to San Pietro and continue northwards on the road towards Ditella, following uphill the sporadic signs to either Calcara or Fumarole. Just northwest of Ditella (location 3), look out for outcrops of the reworked scoria (basaltic-andesite to andesite), rich in euhedral clinopyroxene crystals, millimetres long, erupted from the Punta Torrione complex at some time between 60 000 and 13 000 years ago.

Passing the hamlet of Calcara, a rough track (next to a radio mast) descends from the road to the shore (location 4). At the foot of a raised beach platform, peculiar mud-scoops emit steam from fumaroles that have intensely leached dacitic lava domes, which display well developed radial jointing, and left behind deposits of sulphur. From here, San Pietro can be reached either by retracing the route through Ditella or by following a footpath (marked red-white-red) over the top of the island and then back via Drauto.

Other sites to visit (around the island)
Starting from San Pietro (also Lipari town), day boat trips (including lunch) are available for visiting Panarea, Basilluzzo, Stromboli and Lipari. Contact the Pignataro Shipping Service (090-981.1417, mobile 0368-675975).

Panarea: further reading

Websites
For basic Panarean statistics:
http://www.geo.mtu.edu/~boris/eolie.html
For general tourist information:
http://www.tau.it/aapitme/isoleen.html

Literature
Calanchi, N., C. A. Tranne, F. Lucchini, P. L. Rossi, I. M. Villa 1999. Explanatory notes to the geological map (1:10 000) of Panarea and Basiluzzo islands (Aeolian arc, Italy). *Acta Vulcanologica* **11**, 223–43.
Gabianelli, G., P. Y. Gillot, G. Lanzafame, C. Romagnoli, P. L. Rossi 1990. Tectonic and volcanic evolution of Panarea (Aeolian Islands, Italy). *Marine Geology* **92**, 313–26.

Other islands

Maps. Istituto Geografico Militare topographic maps: 1:50000 sheets 581–586 (Lipari and Salina). 1:25000 topographic maps of all the main islands on a single tourist sheet, published by Trimboli.

Salina

Salina is easily recognized by its distinctive twin peaks, Monte dei Porri (860 m) and Fossa delle Felci (962 m, the highest point of the archipelago). Subaerial activity ceased about 13000 years ago with eruptions of dacitic–rhyolitic pumice from Pollara, on the island's west coast. As of 2000, a museum describing the geological and archaeological history of the island can be visited in Santamarina di Salina (090-984.3021).

Alicudi and Filicudi

The westernmost islands of the group, Alicudi and Filicudi ceased sub-aerial activity, about 28000 and 40000 years ago respectively. Apart from some dacitic lava domes on Filicudi, the products from both islands are dominated by lava flows and pyroclastics with compositions from basalt to andesite.

Chapter 5

Mount Etna

Mount Etna is a giant of a volcano, over 40 km across and more than 3 km high. Little wonder, then, that the mountain has been the abode of gods and giants, whether the home of Vulcan, the prison of Enceladus, or the thunderbolt factory of the Cyclops brothers. It towers over eastern Sicily (Figs 5.1–5.3) where it dominates the lives of more than 2 million people, most relying on the volcano's activity to maintain the fertile soils that support countless vines, olive trees, and orange groves.

Etna is the largest continental volcano in Europe and by far the most active. The summit vents are degassing or exploding almost constantly, and every few years lava eruptions occur from the lower flanks. Recently, major lava effusions have destroyed land and property in 1971, 1981, 1983, and 1991–3 (Table 5.2). The most recent of these eruptions was the largest for over three centuries, extruding between 220 and 300 million m^3 of lava within 15 months. During 1992, lavas threatened the town of Zafferana on the east flank of the volcano, persuading worried officials to build a major earth dam to contain the flows. When this failed, 7 tonnes of explosive were used to divert the lava from its original course. Zafferana was saved, but disagreement still exists about whether or not the lavas would have reached the town without intervention. Following the 1991–3 eruption, magma levels in the volcano remained low and activity subdued, until 1997 when things really began to hot up once again. Since then activity has been confined to the summit region, and particularly the rapidly growing Southeast Crater (in fact a cone). Spectacular lava fountaining up to 1 km or more in height is now a common sight here, accompanied by frequent lava flows that sometimes reach as far as the Valle del Bove, the spectacular depression excavated from the eastern flank of the volcano.

The growth of Mount Etna

Etna has been active for at least 500 000 years, with the earliest eruptions taking the form of basalt flows, pillow lavas and shallow intrusions in a

Figure 5.1 The upper flanks of Mt Etna, looking north from the outskirts of Nicolosi. The white columns signal steam emission from the summit vents (left) and the Southeast Crater (right). The apparent right-hand angular peak is the Montagnola cone, some 4 km closer to the observer than the summit region. Note the gentle overall slopes, reflecting the dominance of lava effusion in the volcano's construction.

coastal environment. The bulk of the mountain has been constructed over the past 100 000 years or so, during which time several major edifices have grown and collapsed. The current volcano, known as Mongibello (a corruption of Monte Jebel, a mixture of Italian and Arabic leading to "mountain of mountains"), is built upon the shoulders of several previously active summits, including Trifoglietto, Vavalaci, and Ellittico. The stratigraphy of Etna is complex, incomplete, and still a source of discussion, so only a simplified stratigraphy and chronology are shown in Table 5.1.

Etna forms part of an extended volcanic front that runs from the Roman and Campanian volcanic provinces in western Italy to the islands of eastern Greece, and which owes its existence to continuing collision between the African and Eurasian plates. Sicily lies close to the junction between the plates, where, because of surface updoming and deformation, the crust has been dissected by several important regional faults. Where these faults intersect, the weakened crust has permitted magma to rise and emerge at the surface, escaping at an average rate of about $1 km^3$ every millennium. The regional influence on Etna's structure is today reflected in the distribution of flank vents, which tend to open along zones cutting the volcano along NE–S and E–W directions (Fig. 5.2).

During historic times, eruptions of Etna have been dominated by

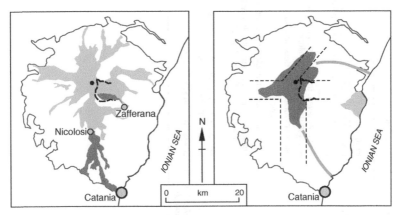

Figure 5.2 Simplified geological (left) and structural (right) maps of Etna. Virtually the whole volcano (thin outline) is covered by lava flows and subsidiary tephra (left). Lavas effused since the sixteenth century are shaded, highlighting (dark shading) those from the 1669 (towards Catania) and 1991–93 (towards Zafferana) eruptions. The Valle del Bove (dashed) lies east of the summit (filled circle).

Etna has grown up at the intersection of regional faults (right). The regional trends are shown by the concentration of vents along a NE–S and more diffuse E–W trend (dashed lines) that both cross the summit region. The dark shading shows the area within which one or more vents occur per km^2. The pale shading on the east coast marks the Chiancone fanglomerate, consisting of material derived from the Valle del Bove. The regional fault directions and eastward tilting of Etna's basement mean that the coastal flanks of the volcano (between grey lines) are slowly slipping into the sea.

Table 5.1 Simplified stratigraphy of Mount Etna.

Centre of activity	Principal formations	Products	Notable events	Approximate age ranges (years)
Recent Mongibello (Il Piano)	Piano del Lago Belvedere Leone	Lavas in historic times; accompanied by thick ashfall deposits and ignimbrites in prehistoric times	Formation of Valle del Bove, Chiancone, Montalto Formation & Biancavilla ignimbrites	<15 000
Ancient Mongibello (Concazze)	Ellittico	Primarily lavas; thick ash sequences accompanying caldera collapse	Formation of Ellittico caldera; formation of extensive debris flows	<35 000–40 000
Cuvighiunni		Mainly pyroclastic fall deposits; some lavas		
Giannicola		Primarily lavas		
Vavalaci		Lavas with some pyroclastic fall deposits	Possible caldera formation	
Trifoglietto		Thick pyroclastic fall deposits plus lavas; possible pyroclastic flow deposits	Phreato-magmatic pyroclastic deposits attributable to eruption through snow and ice cap	60 000–80 000
Ancient alkalic centres	Calanna Paternò volcanic neck	Primarily lavas. Some ashfall, debris flow, and vent agglomerate deposits		90 000–170 000
Basal tholeiitic lavas	Aci Trezza-Aci Castello Adrano–Paternò	Lavas and shallow intrusions		225 000–600 000

Table 5.2   Major flank eruptions on Etna, 1971–93.

Year	Flank	Duration (days)	Volume (10^6 m³)
1971	NE, SE	69	75.0
1974, January	W	17	2.4
1974, March	W	18	2.1
1978, three episodes	E, SE	55	31.0
1979	E	6	7.5
1981	NW	6	18.0
1983	S	131	100.0
1985	S	127	30.0
1986	E, NE	120	60.0
1989	E, NE	12	26.2
1991	E	461	220–300

outpourings of lava, a feature reflected by the shallow slopes of the volcano, which rarely exceed 10° (Fig. 5.1). More than 40 major flank eruptions have been recorded since 1536, of which a sample is given in Table 5.2. The relatively quiet extrusion of lava on the flanks is often accompanied or preceded by episodes of Strombolian activity, involving the ejection of magma as a shower of partially molten fragments with common dimensions from centimetres (scoria) to metres (bombs). These fragments are emplaced ballistically, forming scoria or cinder cones that may extend hundreds of metres in basal diameter and reach heights of 150 m. Over a hundred of these parasitic cones dot the flanks of Etna, marking the positions of eruptive fissures hundreds or thousands of years old.

Strombolian activity is also common, if not dominant, above 3000 m above sea level, where it has helped to fashion the summit complex (3000–3330 m), within which are located three of the four currently active vents. The summit region continually changes over time in both size and shape, sometimes collapsing but then rebuilding itself in succeeding centuries. The summit region is last reported to have collapsed following the huge eruption of 1669, which began near the south-flank town of Nicolosi and destroyed much of Catania. However, it has now built itself back up to form a truncated cone about 2 km across at its base and around 500 m at its top. The summit complex, or Central Crater, contains two open vents, each over 100 m across and of similar depth: La Voragine ("the chasm") and La Bocca Nuova ("the new mouth"). Although the Voragine is a well established vent, the Bocca Nuova opened up in 1968, rather disconcertingly between two groups of tourists! The two craters are buttressed to the northeast by the Northeast Crater, something of a misnomer now as what was a crater when it was formed in 1911 is now a cone, the top of which is (at the time of writing) the highest point of the volcano. Since 1979 a second cone has been growing around the Southeast Crater (born as a pit in 1971), at the southeastern foot of the summit complex, and is now the most active and most prominent feature on the upper southern flank of the volcano.

When not exploding, the summit vents are continually degassing and the smell of sulphur dioxide in the vicinity can sometimes be overpowering, testifying to fresh magma perhaps only a few tens of metres below. On occasion, activity at the summit can be spectacular, with columns of black ash being ejected over 10 km into the atmosphere and falling over an area sufficient to close Catania airport more than 20 km away. Sometimes, as at the Northeast Crater in 1986 and several times over the past few years at the Southeast Crater, gas-rich magma is erupted violently enough to generate continuous fountains of magma that can reach heights of 1 km or more and last for tens of minutes. For magma rising beneath the summit vents to feed flank eruptions, which can occur over a kilometre below, magma must be rapidly transmitted over distances of several kilometres. This is accomplished, when the central conduit is full, by the magma exerting sufficient pressure on the conduit walls to open and propagate along a new fracture. This magma-filled fracture, or dyke, can travel several kilometres in a matter of hours, initiating a flank eruption where it intersects the surface.

Because the slopes of Etna are resurfaced so rapidly (~30 per cent since 1536; Fig. 5.2), the products of older, and especially prehistoric eruptions, are rarely seen. However, there are excellent exposures in the cliff walls surrounding the spectacular Valle del Bove, an 8×5 km slice cut from the eastern flank of the volcano. Although alternative explanations have been proposed, including glaciation and fluvial erosion, it is now generally accepted that the Valle del Bove was formed by gravitational collapse of the east flank of the volcano, perhaps triggered by an eruption. The age of the Valle del Bove remains problematical and proposed dates range from over 20000 years to as little as 5000–6000 years. Recent research using the innovative cosmic-ray-exposure dating technique points to an origin only 3500 years ago. Intriguingly, this date is close to that of an ancient cataclysmic eruption of Etna that resulted in the evacuation of much of eastern Sicily. Perhaps then, this is the first recorded account of a catastrophic lateral collapse of a volcano – à la Mount St Helens in 1980. In fact, Etna remains unstable today, and its growth on a tilted clay-rich basement rock has led to the detachment of the entire east flank from the rest of the volcano. This mobile sector – bounded by the active Pernicana Fault to the north and less well defined fault structures to the south – now creeps seawards at a rate of a centimetre or two a year (Fig. 5.2).

The cliffs surrounding the Valle del Bove, which rise to a kilometre in height along the western wall, provide superb sections through some of the prehistoric edifices, allowing the nature of past eruptive activity to be determined. Although lava flows remain ubiquitous, pyroclastic products dominate many of the exposed sequences, particularly those erupted by the Trifoglietto volcano, which was active – along with neighbouring centres – over 50 000 years ago. It appears that at this time (the height of the last ice age) Etna was covered by a permanent ice cap, leading to violent explosive blasts as magma and water or ice came into contact. Many of the deposits exposed within the Trifoglietto sequences consist of large blocks in a fine ashy matrix, testifying to the large energies involved in their ejection. Although a return to such violent activity is unlikely without an accompanying change in the climate, special circumstances appear to have led to large and damaging explosive eruptions as recently as 122 BC, so the inhabitants of Etna and the surrounding countryside must always be prepared for the unexpected.

The first lava diversion: 1669

Etna's lavas are normally erupted at temperatures close to 1100°C. As they advance, their margins and surfaces solidify, restricting lateral spreading and slowing their rate of advance from typical starting values of 1–10 km per day. Flows thus develop their own channels and tubes, within which hotter lava is transported from the vent to the active front. Such self-structuring has been important to controlling flow behaviour on Etna because, by breaking the solid margins, internal lava can be forced to escape and to flow in another direction. The eruptions of 1669, 1983 and 1991–3 saw three determined attempts to alter a flow's advance. Of these, the 1669 efforts are the earliest on record anywhere and they well illustrate the practical problems of designing defensive measures today.

Effusion began on 11 March 1669, after three days of violent ground shaking around Nicolosi (15 km south of the summit at 700 m above sea level). Lavas poured from the lower end of a fissure system 8 km long and, initially advancing at 4–9 km a day, had within two weeks destroyed at least eight villages up to 10 km from the vent. On 29 March, a new major flow emerged from a breach in an earlier stream and was the first to reach Catania's city wall on 16 April.

As lava invaded the city, houses were demolished immediately ahead of the main fronts, inflammable material was removed and stone barriers were erected across the flow paths. At several locations – presumably those where lava advance was slowest – flow fronts were halted and even diverted. In view of this success, Don Diego Pappalardo and some 50 others from Catania and Pedara were inspired to try and breach the flow near its source at Nicolosi. Protected from the heat by water-soaked goat skins, the group successfully breached the flow margins with makeshift picks and axes, creating a new stream travelling southwest. As news of the enterprise spread, a crowd of 500 from Paternò, a settlement west of

Catania and at risk from the new flow, also climbed to Nicolosi and with menaces forced the first group to desist. As a result, the breach healed and the flow continued towards Catania (Fig. 5.2).

In the end, whole tracts of Catania were destroyed by one of Etna's largest historical effusions (nearly 1 km^3). However, the events showed the possibility of diverting or retarding a flow's advance and provided a blueprint for new attempts some 300 years later. These recent attempts, in 1983 and 1991–3, also involved the erection of barriers and the breaching (by explosives) of flow margins. As in 1669, none of the attempts was an unequivocal success, but each has yielded valuable insights into making future interventions more efficient.

Rock chemistry and petrography

The composition of Etna's historical lavas has been remarkably uniform and constrained between alkali basalts and trachybasalts (hawaiite and mugearite). They are normally porphyritic with some 20–25 per cent by volume plagioclase feldspar, 10 per cent clinopyroxene and 3 per cent olivine as major phenocrysts. The principal petrographic variation concerns the typical size of plagioclase phenocrysts. Lavas erupted during the late sixteenth century and throughout the seventeenth century are characterized by abundant plagioclase phenocrysts about 5 mm in length, giving the lavas the pale speckled appearance known locally as a cicirara texture. Plagioclase phenocrysts in more recent lavas have been generally smaller, reaching 2–3 mm in length. The difference in petrography may reflect that the earlier lavas resided close to the surface for intervals longer than has been common for post-1700 eruptions; if so, then the storage capacity of the shallow magmatic system must be able to evolve over periods of decades to centuries.

Long-term changes in Etna's feeding system are shown by variations in its prehistoric products. Although Etna's basal magmas are tholeiitic basalts, some of the pyroclastic flows and dykes exposed in the walls of the Valle del Bove have trachyandesitic and trachytic compositions. Over hundreds of thousands of years, therefore, Etna's magmatic system has evolved from effusing basaltic lavas, to storing magma long enough to feed trachytic Plinian events, and back again to the outpouring of basalts (alkali, rather than tholeiitic).

The present magmatic system feeding Etna

Historical effusions on Etna have ranged in volume from about 1 million m^3 to $1 km^3$. Although smaller volumes tend to be erupted more frequently, the volumes from flank events cluster around 30–50 million m^3 and 80–120 million m^3. Volumes of a few million cubic metres can be explained as the result of drainage from a central conduit passing through the volcanic edifice to the summit. In contrast, volumes tens of millions of cubic metres or more probably reflect the volumes of magma batches newly entering the volcano from below.

A batch feeding system resembles that inferred for Vesuvius between 1631 and 1944, an analogy strengthened by the restricted petrographical range of each volcano's historic products. Below Etna, the batches appear to be fed from an intermediate reservoir about 20 km below sea level, an ellipsoid 45 km by 60 km wide and 8 km thick across its middle and containing about 14 per cent of molten magma (about $1600 km^3$ of magma in total). The mechanisms constraining batch volumes are a topic of current research, but it is likely that a better understanding of the Etnean system will also improve that of the magmatic system operating beneath Somma–Vesuvius.

Excursions to Etna

Etna can be visited at all times of the year. Snow may restrict access to the summit between December and March, especially during the skiing season from December to February (above about 1800 m on both the north and south flanks). Summer tourism peaks in July and August, during which a good 30–60 minutes should be allowed for queuing to ascend the volcano by cable car or mountain bus.

Since 1987, most of the volcano above 800 m above sea level has been designated a national park (Parco dell'Etna), so that access to vehicles is severely restricted beyond public roads. For the latest information on access, as well as guided tours, call the headquarters of the national park at Nicolosi on 095-914.588.

The upper flanks and the summit

Etna's summit region offers an unprecedented opportunity to witness volcanic activity. Be aware that explosive eruptions from active vents can occur without warning and **at no time should the volcano be considered "safe"**.

Getting to Etna

Etna can be reached directly by road and rail, by air via Catania and by sea via Milazzo.

By air to Catania   Catania airport (Fontanarossa) lies about 10 km south of the city. It is served by direct flights from New York, several European cities (including Frankfurt, Geneva and Dusseldorf) and other Italian airports. Direct services (especially charter flights) run episodically from the UK, but timetables are liable to vary without warning; otherwise, it is necessary to change at an intermediate airport. A bus service (Alibus) connects the airport to Catania's city centre and mainline station (05.00–24.00 h).

By rail to Catania   Regular mainline services connect Catania to all major cities in Italy. Local (and some long-haul) trains also stop at stations along Etna's eastern coastline.

By sea to Milazzo   Long-haul ferries (traghetti) connect Milazzo, on Sicily's northern coast, to Naples via the Aeolian Islands (departures Monday and Thursday; arrivals Wednesday and Saturday). Catania can be reached by train, changing at Messina.

From Catania to Etna   A local railway line, the Circumetnea, runs around the whole volcano, connecting coastal towns as far as Giarre (change of train might be necessary) and then inland via Piedimonte, Linguaglossa, Randazzo (change of train might be necessary), Maletto, Brontë, Adrano, Biancavilla and Paternò (check locally for latest times and connections). Some mainline trains also stop along the coast between Catania and Giardini–Naxos (off the volcano, below Taormina).

Regular bus services (AST, using blue coaches) run from Catania station to Nicolosi on Etna's southern flank. A daily service (departing from Catania at 08.15 h) continues to the cable-car station at the Rifugio Sapienza (leaving for Catania at 16.00 h). Bus services also connect to other towns on the volcano (either directly or after a change; check locally for the latest information). The northern route to Etna's upper flanks begins at Linguaglossa.

Access by car is available from several exits along the coastal motorway A18 and the main road SS114. Inland, the foot of the volcano is traversed by (anticlockwise from south to north) routes SS284 and SS120.

It is especially important to appreciate that significant topographic changes may have occurred in the summit area since it was last visited by the authors. In particular, paths and tracks may have been covered by ash or lava, and the forms of the summit craters may be quite different as a result of eruption or collapse. Meteorological conditions may also vary greatly, even within a few hours, and cloud or plumes of volcanic gas (the latter particularly on cold, damp days) may reduce visibility to a few metres. **Under these conditions the summit craters should on no account be approached**. Latest reports (2001) indicate that recent lava flows from the Southeast Crater are threatening the Torre del Filosofo, so it may be that – at least for a time – buses will stop short of this point and a longer and perhaps off-road walk will be necessary to reach the summit vents.

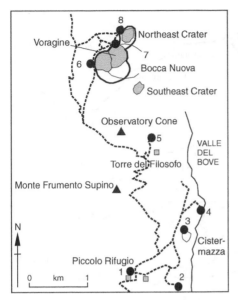

Figure 5.3 Location map for upper-flank itinerary. Numbers refer to localities in the text. From the Piccolo Rifugio, northwest of the top cable-car station (unlabelled filled square), the route continues to Montagnola (2) then north alongside the rim of the Valle del Bove, passing the Cisternazza and, **if access is permitted**, the Torre del Filosofo and the summit craters beyond. Eruptive centres in the summit region are the Bocca Nuova , the Voragine, the Northeast Crater and Southeast Crater. Other major pyroclastic cones are Monte Frumento Supino and the Observatory Cone.

Access to the summit region of Etna is most convenient from the southern side and this is the itinerary followed here (Fig. 5.3). It is also possible, however, to take four-wheel drive buses up the north side of the volcano, and some tourist day trips from Taormina may follow this route. Public roads from Zafferana and Nicolosi (Fig. 5.4) terminate at the small tourist resort of Etna Sud at an altitude of 1923 m, where there is extensive parking for cars and coaches. Overnight accommodation is available at the Hotel Corsaro (with excellent views from the 1983 lava flows; 095-914.122/ 780.9902) and the Rifugio Sapienza (095-916.356) and is advisable if an early morning start and a summit attempt are planned (average prices are about £25.00–30.00 per person per night). From the car-park, a vehicle track covers the remaining 8 km to the summit. Although this is barred to private vehicles, it is possible to walk or mountain bike along the track free

Excursion time Day trip over approximately 12 km.

Maps Istituto Geografico Militare topographic maps: 1:25 000, Sheet 624/I/ Monte Etna. Club Alpino Italiano, 1:60 000 geological map, Mt Etna, Carta naturalistica e turistica.1:60 000.

Difficulty Moderate to difficult, mostly along rough vehicle tracks. Steep slopes of unconsolidated ash and blocks mark the path to the summit area. Walking boots with good ankle support are essential, as are climbing helmets in the vicinity of active craters. A first aid kit is strongly recommended.

Vertical ascent/descent: approximately 2200 m (from the top cable-car station to the summit and back down to Etna Sud).

Figure 5.4 Location map showing the principal roads and towns mentioned in the guide. The A18 motorway (grey) is flanked on its coastal side by the older SS114. The Circumetnea railway runs almost alongside the road loop from Catania, through Giarre, Randazzo, Adrano and Paternò.

of charge. Alternatively, take the cable car (funicolare) or four-wheel drive buses, or both, to either the top cable-car station at La Montagnola (2504 m) or to the Torre del Filosofo (2920 m). Both single and return tickets are available (ranging from about £6 for a single upward trip on the cable car to around £22 for a return trip to the Torre del Filosofo (the latest prices can be obtained from the SITAS company on 095-911.158), offering the opportunity to walk down if time permits. At the time of writing (2001) the guides on the south side of the volcano are not taking parties of tourists to the summit craters, and if you buy a ticket to the Torre del Filosofo the guides will stop you continuing on foot. If you wish to visit the summit, it is necessary, therefore, to walk from the car-park in Etna Sud or from the top cable-car station at La Montagnola. As the first part of the route is the steepest and least interesting, it is convenient to take the cable car to La Montagnola, from which the itinerary starts.

On a clear day, excellent views are afforded from the cable car. Looking south and down hill, many parasitic cinder cones are visible, marking the locations of eruptions on the lower southern flanks. Particularly well displayed is the string of cinder cones known as Monti Silvestri, which span the Etna Sud to Zafferana road and which fed the flank eruption of 1892. Looking north and up hill, the ash and grass covered slope to the right marks the southern edge of the Valle del Bove – a gigantic depression surrounded by steep cliffs that will be viewed at closer quarters later in this itinerary. As the cable car nears the end of its journey, the summit becomes visible. To the left is the summit cone itself, containing two active vents, at least one of which is usually degassing, and to the right the Southeast Crater, currently the most active and rapidly growing vent on the volcano. On arrival, proceed from the cable-car station, which is located on the flanks of Montagnola, a large scoria, ash, and lava cone formed during the eruption of 1763, along an obvious vehicle track to the second derelict building. This is the Piccolo Rifugio ("little refuge"), which served as a restaurant prior to 1985, at which time an eruptive fissure opened directly beneath the building, erupting lava flows that bulldozed through the left side of the structure. Take the track up hill past the right side of the Piccolo Rifugio to view the principal eruptive vents that formed directly behind the building (location 1). It is worth spending a few minutes here to examine the vents and structures of the surrounding lava-flow field. The vents are surrounded by spatter cones or hornitos – steep sided columns of coalesced blobs of fluid magma ("blebs") that were ejected with little violence and therefore accumulated directly around the vent. The hornitos are aligned in a broadly N–S direction marking the position of the subsurface dyke that propagated horizontally from beneath the summit to feed them. Lava flows originating at the vents have spread out to the east of the line of hornitos to form a platform of pahoehoe lava displaying a classical ropy surface texture.

Immediately to the west and directly below the lowest hornito is a superb example of a drained lava channel containing in its upper reaches the final "toes" of lava squeezed from the vent. Cross the bulldozed track beneath the ski lift to visit the main part of the 1985 lava-flow field as it crosses the ash slopes below Monte Frumento Supino. A few minutes spent here will reveal more superb channels floored by smooth flat sheets of lava sometimes known as Roman Roads. These "roads" ooze out of the body of the flow, through what are known as secondary boccas or ephemeral vents. These mark the sites at which molten lava from within the flow field breaks through the overlying crust to flow temporarily at the surface. In other places the still-flowing magma beneath has pushed upwards, usually on encountering an obstacle, to break the crust and open it up like a flower. These push-up structures – known as tumuli because of their resemblance to ancient burial mounds – are typically elongated along the length of the channel and are common to many Etnean lava-flow fields. Between the channels and tumuli, especially where the slope is steeper, slabs of flat and ropy pahoehoe give way to the rubbly lava known as aa. This less spectacular lava surface texture is in fact much more common on Etna than pahoehoe and makes up the bulk of flow field surfaces on the volcano. Within one of the roman roads, a rusty trap door and ladder provide entrance to a lava tube. This can be visited with care, but only if a torch or headlight is available, as there are sudden and dangerous drops in the floor of the tube. Lava tubes constitute the second means, after open surface channels, by which molten lava is fed from the vents to the flow front.

Return to the main vehicle track and continue up hill in a broadly northerly direction towards the Torre del Filosofo and the summit. Alternatively, make a detour eastwards to visit the summit of La Montagnola (easily identified by the cluster of aerials and other hardware on the peak), a cone formed during the 1763 eruption, and to view the interior of the Valle del Bove. Follow a secondary track that heads northwards immediately behind the derelict and rusting cable-car station. Follow the track for about 400 m around the northern flank of La Montagnola and past the obvious flow front of the 1971 lavas on the left. Take the first fork to the right and walk to the top of the ski lift, from which there is an excellent view towards the summit (Fig. 5.5). Although picked over somewhat, the flanks of the La Montagnola cone may still yield some fine volcanic bombs of the fusiform and spindle type in this area. A nearby wooden kiosk with a glass window houses the Etna live-cam that transmits near real-time images of the summit region to the web. Continue up the path on the west side of La Montagnola, which follows the crater rim to its summit (location 2). From here, on a clear day, there is a tremendous view of the Valle del Bove depression, and directly below the cone to the north and east is the main route into the Valle del Bove (see next itinerary). The summit of La Montagnola is used as a base for essential volcano monitoring and telemetry equipment; **do not touch or interfere with this in any way**. Also, be careful not to dislodge rocks that may endanger walkers entering the Valle del Bove beneath your position. Descend the cone by the same route and return cross country to the main vehicle track, which is 0.5 km to the west.

Either take the second fork on the left since leaving the Piccolo Rifugio and follow the main vehicle track up hill towards the Torre del Filosofo and the summit or head northeastwards along the right-hand fork towards the western rim of the Valle del Bove. After about 1 km the track cuts across an aa flow field erupted during activity at the Southeast Crater in September 1989. These flows are unusually thin and lack the obvious

Figure 5.5 The summit region looking NNW from the main track beyond the foot of Montagnola. The Southeast Crater now rises as a large cone (right) on the flanks of the degassing summit region (flat peak to the left). The Torre del Filosofo is in the centre.

steep-sided and levee-bounded channels of the 1971 and 1985 flows. This reflects their formation, which did not involve the more common extrusion of lava from an eruptive fissure. Instead, these flows, known as rootless or clastogenic, formed from the accumulation and flowage down slope of countless molten fragments ejected during energetic lava fountaining. Follow the western edge of the flow to the right (east) of the track to a large crater known as La Cisternazza ("the cistern"; location 3). This is interpreted as an explosion pit formed during the same eruption (1763) that generated La Montagnola, perhaps formed when hot magma came into contact with saturated soil or buried ice. Within the walls of the pit, lavas erupted at or near the current summit can be seen to dip gently eastwards towards the Valle del Bove. Return to the track and cross the 1989 lava field, taking the right-hand (less obvious) fork shortly after the lavas and following this for a few hundred metres. In a small vehicle-turning area marked by a very large accretionary lava ball, take the obvious path to the right (east) and approach the rim of the Valle del Bove. This viewpoint (location 4; the guides often take bus-loads of tourists here, so it can get busy) offers a spectacular view of the 1991–3 lava-flow field, together with its main feeder channel in the ash slopes of the western wall of the depression, below and to the right (south) of the viewpoint. In addition, if clear enough (mornings before 11.00h offer the best chance of a good view), the lava and pyroclastic sequences of the prehistoric Trifoglietto volcano can be observed, dipping radially outwards from a vent originally located in the southwest corner of the depression and which may have approached 3000m above sea level (see next itinerary). Provided that viewing conditions are good, there are also fine views of the post-Trifoglietto dyke swarms in the northern and southern walls of the Valle del Bove, and the "parasitic" cinder cones of Monti Centenari. Return to the main vehicle track and continue to the Torre del Filosofo and the summit.

The Torre del Filosofo ("philosopher's tower") is an unused building that marks the highest altitude (2919m) to which groups of tourists are normally taken by the buses (when the Southeast Crater is active, even this area is off limits). The building is named after a nearby vantage point, said to have been used by the Greek philosopher Empedocles before he hurled himself into the active crater (despite the fact that he seems to have lived his final years in Greece). There is a small cabin here where drinks and light refreshments may be obtained during the day. The cappuccini and the local liqueur – Fuoco del Etna ("fire of Etna") – are both to be recommended on a chilly day. The Torre del Filosofo is located near the spectacular Southeast Crater, the base of which can normally be approached safely. Activity in this part of the summit region started in 1971, when a large collapse pit formed during the eruption. In the late 1970s, magma reached the surface to form a series of vents that erupted lava and ash and which slowly built a cone over the next two decades or so. Since 1997, activity has accelerated noticeably and violent explosive episodes have constructed an impressive cone that dominates the appearance of the summit when viewed from the south. Throughout much of 1999 and, at an increased output rate, the early months of 2000, lava flows oozed from the base of the Southeast Cone, accompanied by lava fountaining and larger explosions that showered ash over the eastern flanks of the volcano. An obvious well worn path, along which the guides take large parties of tourists, approaches the base of the cone (location 5), but access onto the flanks of the cone itself is barred by ropes. This is a sensible precaution as the field of large blocks and bombs (to metres in diameter) surrounding the crater testify to its explosive and often unpredictable behaviour. During periods of intense activity at the Southeast Crater, the roped-off limits may begin well below the Torre del Filosofo. However distant the Southeast Crater appears, **on no account should the ropes be crossed**. Return along the path towards the Torre del Filosofo and turn right, back onto the main vehicle track heading down hill past the Torre del Filosofo. After a short distance the summit track is reached on the right-hand (west) side, marked by an obvious upstanding block of lava. If summit activity is particularly vigorous, this will be roped off and marked with a warning sign. In such circumstances **you approach the summit at your own risk**. The summit track is not always as obvious as the main vehicle track to the Torre del Filosofo and is regularly covered by several centimetres of ash and lapilli because of explosive eruptions at the summit vents. Follow the track for a kilometre or so towards the base of the cone before bearing right along a path heading diagonally up the flanks of the summit cone. This may be more or less distinct, depending on the thickness of the covering of tephra (mainly ash and lapilli) ejected from the summit vents during the preceding weeks and months. The path winds its way, becoming ever steeper towards the northern rim of the Bocca Nuova crater. It is probably best to don protective helmets once on the path.

Always look ahead and upwards, keeping an eye out for explosions from the summit vents. In particular, before climbing the last few tens of metres to the crater rim **spend a few minutes watching for explosions. If these are visible then the vents should not be approached. If no such phenomena are discernible, continue to the crater rim. The round trip to the summit craters should take little more than an hour – do not linger unnecessarily in the crater area.**

The Bocca Nuova (location 6) opened by collapse in 1968 between two groups of tourists. Then it was only a few metres across – now, however, it has been enlarged by collapse to over 100 m in diameter.[1] Much of the

Figure 5.6 The Bocca Nuova in 1999 was a crater some 50 m deep. By August 2000, the outer parts of the crater had been completely infilled by new lavas, some overflowing the western (right) flanks of the summit cone. The ridge running from top to bottom marks the 1999 crater rim. These topographic changes are the greatest for decades at the summit.

crater rim is overhanging and great care should be taken when approaching the edge. Like all three active summit craters, the Bocca Nuova can erupt explosively at any time (as evidenced by large blocks and bombs scattered over the surface) and must, therefore, be treated with respect. In

1 Much of the Bocca Nuova has since been infilled by new lavas (Fig. 5.6).

September 1979, nine tourists were killed and several others seriously injured, following a relatively minor explosion as the vent cleared itself of old material (Ch. 1). In June 1999, the Bocca Nuova was floored by flat-lying lavas, within which three active vents exploded sporadically, hurling molten ejecta to heights of several tens of metres. Such Strombolian activity, although not particularly impressive during daylight hours, provides a fine spectacle as darkness falls.

Head northwards (a path should be visible) to the rim of the largest of the summit craters, La Voragine (location 7). If the volcanic gas is not too dense, interbedded lava and pyroclastic units can be discerned in the crater walls, which also reveal the presence of radial dykes. In June 1999, the Voragine was quiet apart from generating thick fume and deep rumblings suggesting the existence of vents deep within the crater. However, like the Bocca Nuova the Voragine can explode violently. From the rim of the Voragine, cross to the obvious cone of the Northeast Crater (location 8) and climb diagonally the western side of the cone, along a path if discernible. The slope flattens out as the rim of the crater is approached, but great care must be taken, as collapses of the walls of the crater are common. Typically, this is evidenced by gaping arcuate fractures that should on no account be crossed. Although often degassing quietly, the Northeast Crater can erupt violently and hurl large blocks to distances of a kilometre or more. The flank of the Northeast Crater cone affords an excellent view (on a clear day) northwards to the Aeolian Islands. If viewing conditions are ideal, the column of fume emanating from Stromboli should just be discernible.

Retrace steps carefully down hill to the main vehicle track and walk back to the top cable-car station at La Montagnola or continue down the track to Etna Sud. Note that the last cable car down leaves around, or sometimes before, 17.00 h. Be sure to check the exact time before leaving in the morning.

The Valle del Bove

The Valle del Bove ("valley of the ox") is a spectacular cliff-bounded amphitheatre dug out of the eastern flank of Etna (Fig. 5.7). Although its formation and age remain problematical, an origin either partly or entirely attributable to lateral gravitational collapse appears likely. Enclosed on three sides by cliff walls and steep ash slopes up to a kilometre high, the Valle is open towards the east, where it merges gradually into the lower eastern flanks of the volcano. Exposed within the surrounding walls are superb sequences of the explosive and effusive products of at least five of

Excursion time   Day trip over approximately 7 km.

Maps   Istituto Geografico Militare topographic maps: 1:25 000, Sheets 624/I/ Monte Etna and 625/IV/Sant'Alfio. Club Alpino Italiano, 1:60 000 geological map, Mt Etna, Carta naturalistica e turistica, 1:60 000.

Difficulty   Difficult, over steep slopes of unconsolidated ash and grass. A first aid kit and walking boots with good ankle support are essential. Climbing helmets are strongly recommended.

Vertical ascent/descent   Approximately 1000 m.

the prehistoric centres of Etnean activity, and perhaps more. Access to the Valle del Bove is difficult, and care is required at times because of the tricky nature of the terrain, particularly on the steep slopes of unconsolidated ash that offer the easiest entrance.

Since the great eruption of 1991–3 the appearance of the Valle has changed dramatically and access has become considerably more difficult – at least to the floor. Prior to the eruption that started in December 1991, the southern half of the Valle del Bove was floored with lavas primarily erupted during the past two centuries (Fig. 5.7). These had degraded sufficiently to permit the growth of stretches of long pampas-like grass, and were dissected by several ephemeral streams. The distinctive brilliant yellow Etna broom bushes covered much of the lower reaches of the western backwall, and silver birch clustered close to the base of the cliffs that formed the southern edge of the depression. However, all these are now gone, buried under a black lava field up to 250 m thick that today floor the entire southern half of the Valle del Bove. A rock buttress that towered hundreds of metres above the original floor can now just be discerned as a small isolated pillar in the southeast corner of the depression, completely surrounded by the fresh lavas. Before the eruption of the early 1990s an expedition into the Valle would have followed a path down the ash slopes below Montagnola to the floor, followed by a relatively easy walk across the floor to a low point on the southern wall that marks the opening of the Valle degli Zappini ("valley of the shepherds"), and then out. However, because of the difficulties of negotiating the new lavas, the route described in this excursion (Fig. 5.8) will drop close to the new flow field before con-touring up slope again to the rim of the southern wall a few kilometres east of La Montagnola, and then down to a road and back to the starting point.

As for the summit excursion, take the cable car to the station on the western slopes of Montagnola, and follow the vehicle track northwards and up hill past the remains of the Piccolo Rifugio. Turn right along a sec-ond track that runs eastwards immediately behind the derelict cable-car station and follow this for about 700 m along the prominent front of the 1971 lava-flow field. Just before the track splits into two, cross the ash

Figure 5.7 The southwest corner of the Valle del Bove as it appeared before December 1991. This part of the valley floor is now covered by the 1991–3 lava-flow field, which is 250 m thick in places. Outward-dipping strata in the cliff walls are from the prehistoric Trifoglietto eruptive centre.

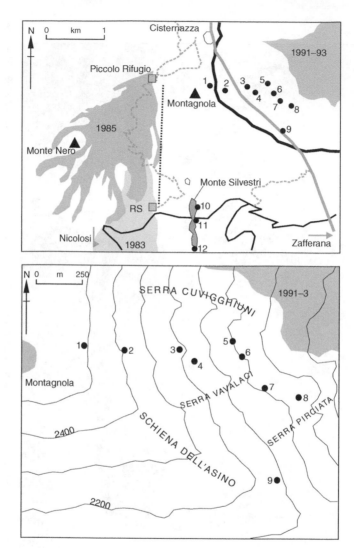

Figure 5.8 Location maps for the Valle del Bove itinerary. Numbers refer to localities in the text. The route (top) begins at the southwestern rim of the Valle del Bove (heavy line) at the base of Montagnola and crosses the trace of the 1989 fracture system (grey line). Localities 10–12 are at the Silvestri Cones (Monte Silvestri), east of the Rifugio Sapienza. Public roads from Nicolisi and Zafferana (solid lines) connect with mountain tracks (dashed lines). A cable car (funivia, dotted line) runs between stations close to the Rifugio Sapienza and Montagnola. Shaded lavas show flows from the 1983, 1985 and 1991–93 eruptions.

Figure 5.9   The entrance to the Valle del Bove. A large mass of yellow volcanic agglomerate marks the only safe route into the valley. **No other route should be taken**.

slopes of the Laghetto and head for the western rim of the Valle del Bove. On nearing the rim, follow the ill defined path southwards along the edge of the depression until a large outcrop of yellow and ochre agglomerate is encountered on the left (location 1). This marks the start of the safe descent into the Valle. **Do not attempt to enter at any other point** (Fig. 5.9). The agglomerate can be viewed at a distance, but the footing can be very tricky at close quarters. The rock forms part of an essentially pyroclastic sequence known as the Cuvigghiuni Formation, which can be better observed at the next location. The agglomerate itself consists of dark and relatively unaltered subrounded blocks of hawaiitic (trachybasaltic) lava and scoria in a strongly altered matrix of finer yellow or yellow-brown material. The appearance of the deposit is reminiscent of hyaloclastite, formed when lava is extruded directly into water, and may reflect contact between lava and snow and ice or water-saturated pyroclastics during the last glacial episode.

Continue past the agglomerate along the poorly defined path that cuts diagonally across the ash slope immediately beneath the summit of Montagnola, and then head eastwards straight down the slope to the first rock buttress (location 2), avoiding the rocky areas on either side where maintaining a foothold is difficult and dislodged blocks can cause problems. Looking back up the slope, outcrops of the Cuvigghiuni formation are exposed on both sides of the ash slope. On the southern side these take the form of outcrops of red welded scoria and yellow agglomerate dipping gently into the slope, and on the northern side by an interbedded sequence of yellow agglomerate and bedded ashes of various hues. Some thin lava flows are also present in the lower part of the sequence. Local unconformities are visible, expressed by the obviously variable dips on the bedded ash sequences. The buttress itself is constructed primarily from hydrothermally altered Cuvigghiuni agglomerate, consisting of an indurated ash matrix containing red scoriaceous fragments and some lithic blocks up to 3 m across. The size of the blocks and the great size range reflect the violent hydromagmatic eruptions that characterized much of the prehistoric activity of Etna during glacial and immediately postglacial times.

Farther to the north, the Cuvigghiuni Formation dips beneath a thick series of lavas that constitute the Belvedere Formation. These flows appear to have ponded in a depression behind the Cuvigghiuni pyroclastic cone and the major prehistoric centre known as Trifoglietto, and are interpreted by some as infilling a caldera formed by the collapse of another major prehistoric centre known as Ellittico. A scar on the face of the cliff formed in the Belvedere lavas may still be visible, representing the source of a recent large rockfall triggered during the opening of the 1989 fracture system. The near-1 km high cliffs that form the backwall of the Valle del Bove are inherently unstable, and dykes propagating parallel to the cliff edge have displaced the entire cliff by up to 5 m eastwards since 1983. This behaviour promotes episodic rock falls and may lead in the future to collapses of the cliff on a larger scale. At the top of the western cliff wall, both the Cuvigghiuni and Belvedere Formations are overlain by thinner eastward-dipping lavas erupted from the current Mongibello centre. Looking down slope towards the floor of the Valle del Bove, many dykes exposed at lower structural levels are obvious, mostly aligned NW–SE. These mark the position of the Southeast Rift Zone, a zone of preferential lateral dyke emplacement that is contributing to the eastward displacement of the east flank of the volcano. Also clearly visible are the Monte Centenari parasitic cinder cones (formed during an eruption in 1852), and the fresh, black flow field emplaced during the major effusive eruption of 1991–3.

The southern wall, representing a surviving remnant of the Trifoglietto centre, is also best viewed from here. Sequences of lavas and pyroclastics, sometimes in excess of 100 m thick, dip radially outwards from the Trifoglietto vent area, which would have occupied a position in the southwest corner of the Valle del Bove. Some authors have proposed recently that the some of the lavas and pyroclastic sequences exposed in the southern wall and allocated to the Trifoglietto centre were in fact erupted by younger centres, including Giannicola (see Table 5.1). Their existence as persistent centres of activity remains, however, unsubstantiated, and the importance of their products (in volumetric terms) is clearly subordinate to that of Trifoglietto. The visible part of the southern wall terminates in the obvious peak of Monte Zoccolaro, which is capped by younger lavas overlying products of the Trifoglietto centre. Beyond and to the north of Zoccolaro, the top of Monte Calanna may just be discerned if the atmosphere is clear, representing the upstanding neck of the pre-Trifoglietto, Calanna eruptive centre, active around 100000 years ago, and constructed from vent agglomerates and other pyroclastic products cut by dykes.

Continue down slope to the south (right, looking down hill) of the buttress where thick (up to 5 m) units of red welded scoria are interbedded with thinner horizons of pale brown hydrothermally altered coarse ash and lapilli belonging to the Cuvigghiuni Formation. Keeping to the centre of the slope, continue down hill towards the next rock buttress, which is formed from a large and obvious dyke segment (location 3). On the right-hand (southern) side of the ash slope, thin aa flows become more common within the Cuvigghiuni Formation, although thick horizons of altered agglomerate, red scoria, ash and lapilli persist. Above and to the south of the second buttress, lavas become dominant, with a succession of 1–2 m-thick aa flows being interbedded with vent-produced pyroclastic material. These flows and pyroclastics constitute the Vavalaci Formation, which underlies the products of Cuvigghiuni and rests upon pyroclastics erupted from the Trifoglietto centre. The lavas dip westwards and southwards off the flanks of Trifoglietto and display dips that are clearly shallower than the underlying pyroclastics, suggesting a discontinuity between the two. Nevertheless, some authors have suggested that the Vavalaci lavas were erupted from the Trifoglietto centre.

After admiring the view from the second buttress, retrace steps for a few tens of metres and drop down onto the next ash slope to the south (right, looking down hill) of the buttress and cross to its southern edge (location 4). Looking back up the slope, thin flows of the Vavalaci Formation are obvious, and to the north the upstanding en echelon segments of several dykes intruded along the Southeast Rift Zone are clearly exposed. Over a hundred dyke segments are exposed in the southwestern corner of the Valle del Bove (Fig. 5.10), some radiating outwards from the Trifoglietto centre, but the majority cutting the products of Trifoglietto activity and defining the obvious NW–SE trending swarm along which the flow of magma has been regularly facilitated during post-Trifoglietto times. Where they reach the edges of the ash slopes, many of the dykes are accessible with care, and they can be examined for a range of structures and textures. Continue eastwards straight down the ash slope (don't be concerned about the slightly disconcerting continued movement of ash, which is close to its angle of rest around your feet). After 100 m or so, note, on the right-hand side of the slope, the steeply dipping sequence of coarse

Figure 5.10 Many dykes forming the Southeast Rift Zone are exposed in the western cliff of the Valle del Bove. Gases in the distance are rising from the 1991–3 eruptive fissure.

pyroclastics and thin lava flows erupted from the Trifoglietto centre and underlying the Vavalaci lavas.

Continue down slope past five obvious dykes on the south side of the slope, all of which may be visited. The dykes are mostly trachytic, with some small plagioclase crystals present and, in some cases, xenoliths of a second more plagioclase-rich rock. Chilled margins may be discerned and good columnar jointing normal to the dyke walls is common. Immediately on passing an obvious lava outcrop in the centre of the ash slope, cross the ash to its northern edge (this can be a struggle in the highly mobile ash). Here (location 5) the crudely bedded and poorly sorted Trifoglietto pyroclastics may be examined. Very coarse, decimetre- to metre-size blocks of lithic and juvenile material in an ash matrix are interpreted as evidence of violent hydromagmatic activity involving the explosive interaction of rising magma and groundwater or water-saturated pyroclastics. Just down slope, three NNW–SSE trending post-Trifoglietto dyke segments may be examined; looking back up slope at this point, two very thick Trifoglietto dykes can be seen cutting the pyroclastic sequence and orientated in a broadly E–W direction. These dykes would have propagated from the central conduit system of the Trifoglietto volcano located 1.5 km or so farther east. Compositionally and petrographically these dykes differ significantly from the post-Trifoglietto dykes. They are pale mugearites (trachybasalt) characterized by obvious black or very dark green pyroxene phenocrysts that may be up to 3 cm in length.

Cross the ash slope again to the south side to the higher of two steeply dipping (eastward) post-Trifoglietto dykes. This strongly porphyritic dyke contains obvious plagioclase phenocrysts in a dark groundmass. Walk down slope to the lower dyke (location 6), which shows a clear segment offset. From the grassy slope below the dyke, a clear view of the eastern dyke wall can be had. Grooves (flow lineations) dipping at 30° towards the northwest indicate the source in this direction. Also, circular depressions in the dyke wall represent outstanding blocks of country rock around which the magma has flowed during emplacement. Columnar jointing perpendicular to the length of the dyke is also well displayed. Petrographically, the dyke is trachytic, containing sparse, small plagioclase crystals and displaying obviously aligned vesicles that support a subhorizontal flow direction. Glassy chilled margins are well developed. Continue up the grassy slope below the dyke, following along its length. Climb down onto a narrow ash and boulder slope and up another grassy slope to a cliff in which a distinctive pale Trifoglietto pyroclastic unit is exposed (location 7).

This deposit contains pale pumice clasts, many flattened or altered to a soft clay-like material, or both, together with indistinct laminations, and have been interpreted as wet pyroclastic flow or surge deposits erupted at a time when the Trifoglietto centre had a substantial cover of snow and ice. The sequence also contains dense dark clasts, which, on closer inspection, prove to be plagioclase-and-kaersutite (amphibole) cumulate material ripped from the margins of the Trifoglietto magma reservoir by the violence of the eruption and incorporated into the flow or surge deposits. Drop down to the next ash slope, partly covered by slabs of indurated pyroclastics (an optional downslope detour may be taken here to examine the surface features of the 1991–3 lava-flow field, although the walk back up hill is a struggle) and cross a sometimes-visible path to the steep grassy slope opposite. Head for a rock buttress with a large obvious rock balancing on top. Climb the grassy slope in front of the buttress to view a dyke and a sill emplaced within Trifoglietto pyroclastics (location 8). The sill lacks the very rubbly base and top of aa lava flows, and is unconformable with the broad dip of the pyroclastic units it intrudes. Continue to the south up the steep, grassy slope in front of the buttress, and look northwest to view the dykes of the Southeast Rift Zone and the main eruptive vent of the 1991–3 eruption (marked by obvious white sublimate) in the western cliff wall (Fig. 5.10).

Head southeastwards diagonally up slope towards the southern rim of the Valle del Bove (a crude path is discernible in places). This route will intersect, at some point, an obvious fault plane that marks the southern edge of a graben-like fracture system formed in October 1989 (location 9). The fractures propagate for over 6 km from the Southeast Crater to just below the Etna Sud to Zafferana road, and facilitated – in the western wall of the Valle del Bove – the passage of lava that fed the 1991–3 eruption. In the southern wall, the southern graben fault shows a downthrow to the south of up to 2 m in places. Follow the fault, which has become a reasonably well defined path, to the rim of the southern wall and head directly down slope (the terrain can be quite rough, so care must be taken) until an obvious path is reached. Follow this down to the main road, turn right and head back up hill towards Etna Sud. The road cuts through a small saddle in a line of pyroclastic cones known collectively as Monti Silvestri (Fig. 5.11), which mark the site of the 1892 flank eruption. The cones are arranged along two fissures fed by dykes propagating horizontally or subhorizontally from rising magma beneath the summit region. Dyke emplacement was recorded by 20 h of seismic activity that culminated in the opening in July 1892 of a 3 km-long fissure system extending from just below La Montagnola (2600 m above sea level) down to an altitude of 1750 m. The first lavas were erupted from vents on the western fissure, but activity rapidly transferred to the second fissure 250 m to the east, where explosive activity led to the construction of Monte Silvestri Superiore,

Figure 5.11 Looking south over the 1892 Silvestri cones (Superiore left, Inferiore, right). The eruption was fed by a dyke propagating laterally from Etna's central conduit. Associated lava flows (down slope from the cones) almost reached Zafferana.

Monte Silvestri Inferiore and Monte Nero, from spatter, scoria, lapilli and ash. Lava production continued until the end of December, forming flows that travelled eastwards for 6 km to threaten the village of Zafferana.

From the small restaurant on the right (north) side of the road, a path climbs around the western side of M. Silvestri Superiore to the highest point (location 10). From here it can be seen that the cone is elongated and roughly aligned N–S along the eruptive fissure. It is constructed from explosive activity at four vents, the positions of which, despite post-eruption infilling, can still be discerned – each separated by a "saddle" of spatter. Exposure within the crater walls is generally poor, although at the southern end there is a sequence of intermixed red scoria and spatter. At the highest point on the southern rim, upstanding outcrops consist of hydrothermally altered scoria and lapilli with some spatter. This material is crudely stratified and coated with brightly coloured sulphur-rich sublimate attributable to fumarolic activity along the crater rim. Just below the eastern and southeastern rim there are obvious upstanding masses of stratified coarse ash, lapilli, and scoria that has been cemented by fumarolic activity and exposed by the erosion of less well consolidated material. In the loose scoria mantling the cone, small but prominent pale xenoliths of sandstone (and more rarely limestone) may be found. These inclusions are particularly common in the ejects from the 1892 eruption and indicate that the feeder dyke must have propagated at least partly within underlying sedimentary basement, which here is probably only a few metres beneath the surface. The xenoliths have been recrystallized and partly melted, resulting in a distinctive web-like texture (viewed under a microscope) characteristic of a rock called buchite, in which surviving and corroded quartz grains are enclosed in a matrix of silica-rich glass. Return via the eastern path to the road and cross to M. Silvestri Inferiore, which is located next to the Ristorante Cratere Silvestri. The flanks of this cone enclose two principal craters, the best exposed of which is immediately adjacent to the road (location 11). Exposed sections of welded spatter are obvious in the near-vertical walls of the vent, overlain by more scoriaceous material on the eastern rim. There is also some ventward-dipping spatter plastered against the inside of the crater walls and representing the waning stages of activity at the vent. A well worn path continues southwards, descending into the floor of the second crater. This is poorly exposed internally, but some spatter can be seen around two vents at the base of the crater.

Walk around the crater rim to the southern end and follow the path down slope across outward-dipping, fumarolically altered, and stratified coarse ash and lapilli. Continue through well exposed spatter ramparts and down towards a large hole that represents a partially collapsed lava tube. Collapse has produced a "window" through which two effusive vents or boccas can be seen. These are the feeders for the 1892 lavas that extend for around 6 km southwards and eastwards towards Zafferana. From the lava vents, climb the path to the east that leads to the rim of Monte Nero, the southernmost of the 1892 cones (location 12). This cone is similar to the northernmost crater of Monte Silvestri Inferiore, with vertical or near-vertical walls plastered with inward-dipping spatter. A few metres above the floor of the crater, an obvious "tidemark" indicates the highest level reached by lava in the vent. Monte Nero appears to rest, at least partly, upon lavas erupted from the boccas to the northwest, and provides from its highest point an excellent view of the surface morphology of the 1892 flow field. Immediately below the southeast rim of the crater, three tumuli are particularly well developed. On a clear day, the crater rim also offers a good view southwards and southwestwards over the Southeast Rift Zone and the highest concentration of cones associated with both prehistoric and historic flank eruptions.

The lower flanks and the coast

This excursion (Fig. 5.12) requires an independent vehicle and focuses on the geology and volcanology of the lower flanks of the volcano. Locations may be visited independently or be combined to form a round trip that can

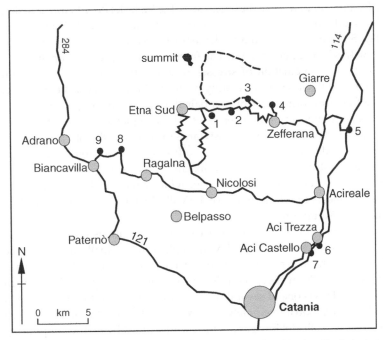

Figure 5.12 Location map for the lower-flank itinerary. Numbers refer to localities in the text. Roads (solid lines) and towns as in Figure 5.3.

be completed in a day. In the latter case, the suggested order of localities, assuming a start at the bottom cable-car station at Etna Sud or at Nicolosi proper, is:

1. the 1989 fractures
2. the Valle degli Zappini
3. Monte Zoccolaro
4. the 1992 lava flow front
5. the Chiancone at Praiola
6. lava columns and pillows at Aci Trezza and Aci Castello
7. the Montalto Formation and the Biancavilla ignimbrites.

From the cable-car station at Etna Sud, head down the road towards Zafferana (or if coming up from Nicolosi, turn right at junction with the Zafferana road). After three tight hairpin bends, the road straightens out and passes a high wall on the left that is broken by two obvious fractures (Fig. 5.12; location 1). These represent the distal end of a 6 km long fracture system that opened in 1989 during an eruption at the Southeast Crater. The fractures define a graben structure (note: the central block of the wall is downfaulted) sourced at the Southeast Crater, which changes orientation

145

to become subparallel to the backwall of the Valle del Bove before crossing the southern wall of the depression and cutting the Zafferana road (Fig. 5.8). To the right of the road, the fractures can be traced for a few tens of metres before petering out. The fracture system probably opened as a result of a combination of stresses resulting from rising magma from below and the inherent instability of the backwall of the Valle del Bove. The fractures form one arm of a pair that opened during the Southeast Crater eruption in September 1989. The complementary fractures are shorter and propagated northeastwards from the Southeast Crater. While lavas issued from these fractures, no effusion occurred from the southern fracture system, which remained inactive. However, this was not to last and in December 1991 the fractures provided a route to the surface, on the inner slopes of the western wall of the Valle del Bove, for the lavas feeding the 1991–3 eruption.

Continue down the Zafferana road until it flattens out temporarily while crossing the 1792 lava-flow field. Park in the small car-park on the left side of the road next to the tree-covered cinder cone of Monte Monaco (location 2). From here, there is an excellent view up the Valle degli Zappini. This is one of several wide valleys (including the steeper-sided Valle del Tripodo immediately to the west) that dissect the outer southern wall of the Valle del Bove and which were formed by meltwater pouring down from the ice- and snowfields that capped the Trifoglietto volcano. Like all fluvially cut valleys, these would have narrowed towards their source. However, the upper reaches of the valleys are no longer visible, having been truncated by the collapse that generated the Valle del Bove. The water runoff must have been spectacular at times, with water pouring over an exposed dyke that forms an impressive cliff at the base of the valley to form a towering waterfall. The dyke still acts as an aquifer trap and a small spring behind it provides much needed refreshment for the sheep and goats that roam the area. If there is time, the top of the dyke can be visited via a series of steps that have been built recently from the valley floor. Recent dates determined using the innovative cosmic-ray exposure dating method indicate that the source of the water that formed the valleys and waterfall was cut off around 3500 years ago, suggesting that the Valle del Bove, or at least its southern part, was formed at this time.

Excursion time   One day or less, according to number of localities visited.

Maps   Istituto Geografico Militare topographic maps: 1:25 000, Sheets 624/II/ Adrano, 625/I/Giarre 625/II/Acireale, 625/III/Aci Catena, 625/IV/Sant'Alfio and 634/IV/Catania Nord. Club Alpino Italiano, 1:60 000 geological map, Mt Etna, Carta naturalistica e turistica. 1:60 000.

Difficulty   Easy.

Return to the road and continue to head down hill towards Zafferana. Looking back up hill during the descent around several hairpin bends, excellent views are afforded (if clear) of the summit and the backwall of the Valle del Bove. After half a dozen or so tight bends, take the obvious exit to the left signposted for Monte Pomiciaro. Follow the winding road up hill until it ends at a small car-park overlooking the Val di Calanna, a secondary depression at the southeast end of the Valle del Bove. The location provides an excellent view of the 1991–3 lava-flow field, which now fill the depression. These flows ponded behind an earth barrier constructed along the open eastern edge of the Val di Calanna to try and prevent the lavas from reaching Zafferana. However, the barrier was overtopped in April 1992 and is no longer visible. The upstanding mass of Monte Calanna represents an old volcanic neck, constructed from volcanic agglomerate and dissected by many dykes, that was an active eruptive centre during the early history of the volcano and is probably over 100 000 years old. On the western side of the car-park, a footpath leads up hill along the southern edge of the Valle del Bove to the peak of Monte Zoccolaro (location 3), where, on a clear day, there are fantastic views across the Valle del Bove and of the summit. The climb to Monte Zoccolaro takes about 30 min and can be tiring on a hot day. Nevertheless, the tremendous view makes the effort very worthwhile.

Follow the Pomiciaro road back to the Zafferana road and turn left. The road traverses a steep scarp slope via a series of tight hairpin bends and enters the outskirts of Zafferana. Immediately on entering the main part of the town, take the first turn on the left in the direction of the village of Ballo (the 1992 lava flow front may also be signposted). Cross a dry ephemeral stream bed and head left up hill. Park at the obvious shrine to the Madonna, which marks the farthest extent of the 1991–3 lava-flow field (location 4). A rough path leads over the flows to a destroyed farm out-building on which the words "Grazie Governo" ("Thank you, Government") are scrawled – a sarcastic comment from a local resident who blamed the damage caused by the lavas on the authorities. A few tens of metres up slope, the remains of a small earth barrier can be discerned through the lava. When the holding barrier was overtopped in the Val di Calanna, several smaller barriers were hurriedly constructed to slow down the flows. However, all were easily overtopped, allowing the flows to destroy several small outbuildings in the area. The flows fortunately stopped before the main part of Zafferana was reached and the explosive diversion of the main channel feeding the flow in the upper reaches of the Valle del Bove ensured that no further flows approached the town. Zafferana sports some pleasant bars and cafes around the piazza, where lunch can be had. Alternatively, head down hill and towards the coast.

Figure 5.13 The Chiancone fanglomerate at Praiola consists of flash-flood and debris-flow deposits. The sequence contains cobbles and boulders originating from the Valle del Bove. Borehole studies suggest that the deposit may be as much as 300 m thick.

Make for Santa Venerina and from here continue east under the autostrada to Trepunti. Here, turn right onto the SS114 and after a few kilometres take another right turn at San Leonardello. Follow signs for Carruba and turn right at the crossroads in the town. After about 1 km turn left for Praiola (easy to miss) and continue to the road's end. Parking in the road is possible, although the area can be busy on weekends in summer. Walk onto the cobble beach and head right (south) for the obvious cliffs (location 5). These are made up of flash-flood and mudflow deposits that belong to the fanglomerate sequence known as the Chiancone. These deposits form a triangular fan directly opposite the open end of the Valle del Bove, suggesting some link between the two, and there is little doubt that the majority of the blocks within the Chiancone have been washed down from the depression. The thickness of the Chiancone (up to 450 m in places, according to borehole data) suggests that it is filling a deep depression, the nature of which is unknown, but which may represent the scar of an earlier lateral collapse event. Coastal bathymetry indicates that the Chiancone fan extends out into the Ionian Sea for several kilometres, although little is known in detail of the form of the submarine extension of the fan. In the cliffs at Praiola, only the top 30 m or so of the Chiancone sequence is exposed, but this is sufficient to provide information on the depositional environment. Large boulders and cobbles, some in excess of 1 m across, testify to a high-energy regime (Fig. 5.13), with normal grading and localized cross bedding indicating that water was the principal transporting medium. Some matrix-rich units showing reverse grading represent debris flows, and occasional fine ash horizons testify to periodic tephra deposition during explosive eruptions. Clast petrology and chemistry indicate that they are representative of the ancient volcanic centres, the products of which are now exposed in the walls of the Valle del Bove. Given that the likely origin of the Valle del Bove was by lateral collapse initiating one or more rock avalanches, it might seem surprising that none of this material is obvious in the Chiancone sequence. However, it is likely that the true rock avalanche deposits are buried at depth, with the material forming the upper part of the Chiancone representing the secondary reworking of this material down slope. At the northern edge of the cliff, however, there is a large boulder on the beach that bears some resemblance to rock avalanche debris. The boulder contains clasts of several different lithologies, which are angular and show jigsaw textures in places, where they have broken apart during transport.

From Praiola, return to the SS114 and follow the road southwards along the coast. A few kilometres north of Acireale the road climbs spectacularly up the face of the Santa Tecla Fault scarp. This marks the onshore extension of the submarine Malta Escarpment, one of the principal structures controlling the ascent of magma at Mount Etna. Take the Aci Trezza exit and park in the town. Walk to the sea front, where superb basalt columns are exposed on the foreshore (location 6). These represent slowly cooled shallow sills intruded into clay-rich deposits during the early history of the volcano around half a million years ago. On the Isole Ciclopi (Cyclops Islands), just offshore, the relationship between the dark basalt sills and the pale grey clays can be easily seen. From Aci Trezza, return to the SS114 before taking the turning for Aci Castello after about a kilometre or so. Park in the town and head for the Norman keep, perched precariously on a rock promontory (location 7). From the viewpoint adjacent to the castle, it is obvious that the promontory comprises spectacular pillow lavas formed by the submarine eruption of basalt (Fig. 5.14). The larger pillows display radial columnar jointing (entablature) reflecting fracture-pattern development during slow cooling. Walk down the stairs to the right of the castle to reach a flat rock platform and head for the base of the rock outcrop beneath the castle. Here, the strata appear to be near vertical, with pillow-rich sequences cut by layers of red-brown hyaloclastite. This material is made up of comminuted basalt resulting from the rapid chilling and spalling of glassy material from molten basalt as it enters the sea. The fine glassy clasts are rapidly altered by reaction with sea water and break down to form the clay mineral palagonite. Looking back towards the stairs, an uplifted marine abrasion platform is obvious on which a much

Figure 5.14 Superb sequences of pillow lavas below the castle and along the foreshore at Aci Castello. Individual pillows show radial cooling cracks, vesicles often infilled by the white feldspathoid analcite, and glassy rims that might be detached from the bulk of the pillow.

younger lava flow rests unconformably. Continue walking around the base of the outcrop to its northern side, where superb cross sections of pillows are exposed, cut in places by basalt dykes. The pillows are vesicular, with the vesicles containing the milky-white mineral analcite. This sodium-rich mineral, related to the feldspars, has been precipitated in the gas cavities through reaction between the basalt and sea water. Individual pillows also display a fine-grain glassy margin that in many cases has broken off and collected in the gaps between adjacent pillows. Continue crossing the pillows and take the stairs that wind up the north side of the castle.

Return to the SS114 and travel north. At Acireale, head west to Nicolosi following a route that takes in Aci Catena, Aci Sant'Antonio, Viagrande, Trecastagni and Pedara. Continue through Nicolosi towards Ragalna Ovest. In the village turn right towards Biancavilla and after about 2 km turn right again at a crossroads onto a minor road heading north. After 1.5 km take a left turn and park near a rubbish dump (on the right) beneath a line of low cliffs that mark the western margin of the Vallone di Licodia (location 8; Fig. 5.15). Exposed in the cliffs are the little-studied deposits of the Montalto Formation, which are interpreted as pyroclastic flow deposits, possibly related to an ancient phase of caldera collapse in the summit region of Etna. Pyroclastic flows are rare on Etna but not unknown. They tend to form following episodes of prolonged magma storage that lead to changes in magma chemistry and the eruption of much more Si-rich,

150

Figure 5.15 The Montalto Formation is believed to be an ignimbrite (pumiceous pyroclastic flow) deposit formed during prehistoric caldera collapse at the summit.

viscous, and therefore violent, explosive eruptions. The deposit here consists of a friable red-brown matrix containing a range of clasts but dominated by fragments of black pumiceous material. The absence of any obvious stratification and the generally chaotic nature of the assemblage indicates that this is not a tephra deposit but has been emplaced by flow. The age of the rocks deposited is not well constrained, but they are believed to have been erupted around 15 000 years ago.

At a T-junction just past the rubbish dump, turn left and continue westwards towards Biancavilla. After 2.5 km turn left again at a second T-junction and cross a small gorge. Note fragmental rocks exposed in the roadside on the right and pull over after a few hundred metres into a layby next to an area of waste ground (if you reach another road you have gone too far). Enter the waste ground and head for the low cliffs that mark the western side of the gorge (location 9). From the top of the cliffs there is an excellent view within the cliff walls of the gorge of the Biancavilla Ignimbrites. Probably erupted at around about the same time as the rocks of the Montalto Formation, the Biancavilla Ignimbrites are pumice-rich pyroclastic flow deposits that in the gorge show crude columnar jointing. This is not uncommon in ignimbrite sequences and it reflects slow cooling following deposition. The ignimbrites are particularly siliceous for Etna, containing over 60 per cent SiO_2, compared with less than 50 per cent for historically erupted basalt lavas. This high silica content would have ensured that eruptions at the time were gas-rich and unusually violent for

Etna, leading to the collapse of explosive eruption columns and the consequent formation of pumice-rich pyroclastic flows that travelled for distances of up to 15 km. Even viewed from the other side of the gorge, some dark, glassy fragments are visible. Closer examination shows that some of these glassy fragments are stretched out. This indicates that the temperatures during and immediately following deposition were high enough for clasts to behave in a plastic manner. Although not well exposed, the top of the Biancavilla Ignimbrite can be examined around the margins of the waste ground and along the side of the adjacent road. No obvious stratification is evident and the deposit is largely homogeneous, containing clasts primarily of grey and red pumice with less common basaltic glass (tachylite) fragments. Rare occurrences of carbonized wood, and even entire tree trunks, have also been reported.

Return to hotel or hostel for a well earned rest and aperitif. Discuss the day's observations over a large bowl of pasta and ponder on the many and varied geological sights and sounds of one of the world's great volcanoes.

Other sites to visit

The northern flanks of Etna

The main itineraries have focused on Etna's southern flanks, since these are the most accessible by private and public transport, and also on foot. The second main route up the volcano leaves from Linguaglossa, past the Rifugio Brunek along the Piano Pernicana, to the Rifugio Nord-Est at Piano Provenzana, the base tourist complex for skiing and from where mountain buses can be taken to the summit region (volcanic activity permitting). More information can be obtained from the STAR company on 095-643.180.

Further reading

Websites
For the latest news on Etna, including "livecam" video of the summit region:
http://www.geo.mtu.edu/~boris/
http://www.iiv.ct.cnr.it
http://www.poseidon.nti.it
For tourist information:
http://www.apt.catania.it

Literature
Barberi F., M. L. Carapezza, M. Valenza, L. Villari 1993. The control of lava flow

during the 1991–1992 eruption of Mt Etna. *Journal of Volcanology and Geothermal Research* **56**, 1–34.

Chester, D., A. M. Duncan, J. E. Guest, C. R. J. Kilburn 1985. *Mount Etna: anatomy of an active volcano.* London: Chapman & Hall.

Gillot P. Y. et al. 1994. The evolution of Mount Etna in the light of potassium–argon dating. *Acta Vulcanologica* **5**, 81–7.

Gravestock, P. J. & W. J. McGuire (eds) 1996. *Etna: fifteen years on.* Cheltenham & Gloucester College of Higher Education, Cheltenham.

Hughes, J. W., J. E. Guest, A. M. Duncan 1990. Changing styles of effusive eruption on Mount Etna since AD 1600. In *Magma transport and storage*, M. P. Ryan (ed.), 385–406. Chichester: John Wiley.

Kilburn, C. R. J. & W. J. McGuire 1997. Mount Etna: its evolution and hazards. In *Volcanism and archaeology in the Mediterranean area*, R. Cortini & B. De Vivo (ed.), 149–62. Research Signpost, Trivandrum, India.

McGuire, W. J. 1990. The rise and fall of Mount Etna. *New Scientist* **125**(1706), 50–54.

Romano, R. (ed.) 1982. *Mount Etna volcano. Memorie della Società Geologica Italiana* **23**.

Glossary

aa lava Hawaiian term for lava flows with extremely uneven surfaces, usually covered by irregular spinose fragments of broken crust that are rarely more than tens of centimetres across. They form lava channels that may evolve into tubes. Their fronts can advance at 10–20 km an hour, but more usually at 1–10 km a day.

agglomerate Coarse pyroclastic deposit containing bombs. Agglomerates are produced during eruption, unlike reworked conglomerates.

aphyric texture A texture describing a rock consisting only of glass or a fine-grain matrix. Phenocrysts are absent. Compare with porphyritic texture.

ash Particles of magma less than 2 mm across.

blocky lava Lava flows with fractured surfaces, usually covered by debris up to metres across. Unlike aa debris, blocky surface fragments are normally angular with smooth sheer faces. The flows typically advance at rates slower than 1 km a day.

bocca Volcanic vent. Ephemeral or secondary boccas are breaches in the margins of lava flows through which hot mobile lava has escaped. The name is derived from the Italian for "mouth" (the Spanish equivalent boca is also used in English).

bomb Rounded volcanic fragment larger than 6.4 cm across. During flight through the air, bombs may develop distinctive fluidally shaped exteriors (fusiform or spindle bombs). The sur-

faces of more viscous bombs may form a network of cracks resembling crusty bread (breadcrust bombs).

boulder Generic term for rounded rock fragments, not necessarily volcanic, more than 25.6 cm across.

bradyseismic activity Slow oscillations of a caldera floor following major caldera collapse. The movements cause very low-level seismicity, most of which is not felt but only registered by instruments. The term was introduced to describe movements in Campi Flegrei, presumed to have begun after the Tufo Giallo Napoletano caldera was formed about 12 000 years ago.

breccia Deposits of mostly angular blocks within a matrix of fine particles. Volcanic breccias normally consist of broken pyroclastic rocks.

caldera A giant volcanic crater (notionally larger than 1 km across) formed by collapse or explosion, collapse being more important among larger calderas. The name comes from the Caldera Taburiente on La Palma in the Canary Islands.

cinder Generic term for coarse volcanic ash and lapilli.

clastogenic See rootless lava flow.

cobble Generic term for rounded rock fragments, not necessarily volcanic, between 64 and 25.6 cm across.

columnar jointing Pattern of fractures produced by the slow cooling of a magmatic body. The fractures develop as a collection of hexagonal (or quasi-hexagonal) columns extending inwards from the cooling surface. When the magmatic body has a strongly curved surface (e.g. lava pillow or lava dome), the columns form a radial pattern about its centre. Extensive joint systems may consist of regular columns (forming a colonnade) interrupted by layers of irregular fractures (forming an entablature).

column collapse See Plinian eruption.

cone Conical constructs built up by the accumulation of material around a vent. They may consist of tephra or a mixture of tephra and lava flows. Cones may be the result of a single eruption (e.g. Monte Nuovo in Campi Flegrei) or the product of extended periods of activity (e.g. the Southeast Cone on Mount Etna). The flanks of some large volcanoes, such as Etna, are dotted by hundreds of small parasitic or adventive cones, which mark zones of weakness in the volcanic edifice.

conglomerate Coarse reworked deposit of rock fragments, not necessarily volcanic, containing cobbles. Reworking is caused by sedimentary processes, from landsliding to river transport. Compare with agglomerate.

crater A pit or depression, typically located around a vent. Craters may be formed during construction of an enclosing cone, by the excavation of rock during volcanic explosions, or by the collapse of ground left without support after magma has been erupted. Craters wider than about 1 km are normally termed calderas.

cross bedding Bedding structure produced when particles in pyroclastic flows and surges settle through the volcanic gas to collect on the ground and move forward as migrating ripples or dunes. In cross section, the particle layers cross-cut each other to produce a pattern resembling broken chevrons.

dyke Vertical fractures filled with solidified magma. When magma stops flowing through a dyke, it solidifies to form a "wall" of volcanic rock that may

be exposed by future erosion. Similar features lying almost horizontally are termed sills.

entablature See columnar jointing.

eutaxitic texture Texture produced by masses of flattened pumice welded together by their own heat.

exsolution Normally applied to magmatic degassing, exsolution describes the process by which volatile constituents (especially H_2O) dissolved in a magma come out of solution to form gaseous bubbles.

fanglomerate A fan-shape deposit of conglomerate. A giant example is the Chiancone deposit at the seaward mouth of the Valle del Bove on Etna.

fissure A surface fracture. Often the surface expressions of dykes, fissures may also open near the rims of unstable slopes, including craters.

fumarole Hole in the surface allowing the escape of volcanic gases and heated groundwater.

hornito Small cone (to tens of metres tall) produced by weak explosivity and usually composed of layers of spatter. The name is derived from the Spanish for a "small oven".

hyaloclastite Fine-grain angular debris produced by the chilling and fragmentation of lava crusts as a flow advances under water.

hydromagmatic eruption An eruption whose explosivity is significantly enhanced by steam from non-volcanic water (e.g. groundwater, lake water and sea water) that has come into contact with magma. Previously termed phreatomagmatic.

ignimbrite A pumice-rich pyroclastic flow, normally associated with, but not exclusive to, large-volume explosive eruptions.

lapilli Magmatic fragments 2–6.4 cm across. Accretionary lapilli or pisolites are produced in eruption clouds when coatings of ash form concentric layers around a tiny nucleus.

lava Magma that has breached the surface.

lava channel See lava flow.

lava dome When lava is extruded onto a near-horizontal surface, it tends to pile up around the vent and then flow away equally in all directions to produce a dome. The dome may grow by intrusion of new lava into the dome interior (endogenous growth) or by the overlapping of many small lava tongues or flows that can escape through breaches in the dome's surface (exogenous growth).

lava flow Magma extruded at the surface as a fluid and able to flow away from its source. Lavas solidify during advance and soon form lateral margins or levées that constrain the hot, mobile material within either open channels or tubes. Solidification also encourages active flow fronts to decelerate, so allowing lava to accumulate within the feeding channel or tube. The accumulating lava may force a breach (ephemeral or secondary bocca) in the lateral margins to propagate a new lava flow. A flow field is the collection of flows produced by a single eruption. Three major types of lava flow – aa, blocky and pahoehoe – can be distinguished by their morphology and preferred styles of emplacement. See aa lava, blocky lava and pahoehoe lava.

lava fountain Jets of magmatic fragments (from bombs to ash) that may reach heights of several hundred metres. They may occur on their own or as the basal jets feeding fragments to buoyant Plinian eruption columns.

lava tube See lava flow.

levée See lava flow.

lithic material Fragments, usually angular, of rock stripped from conduit walls during eruption. Although associated with country rock existing before an eruption, the fragments may also include magma from the current eruption that has been chilled against

the conduit walls and later torn away.

magma Generic term describing all molten rock. For common rock compositions, eruption temperatures are between 900°C and 1200°C, lower temperatures occurring among more-evolved magmas that have greater proportions of the alkali elements sodium and potassium.

meteoric water Surface water (provided by rainfall, snowmelt, lakes and the sea) circulating in the crust. The term distinguishes such water from that derived directly from H_2O escaping from degassing magma.

nuée ardente Strictly a pyroclastic flow of poorly vesiculated magma (from the French for a "hot cloud"), although the term is often used loosely as an alternative to all types of pyroclastic flow.

pahoehoe lava Hawaiian term for lava flows with smooth, continuous surfaces that may locally appear crumpled into a collection of tight folds resembling part of a coiled rope (ropy pahoehoe). Lava is normally transported along tubes from the vent to the front, which oozes forwards as a series of small overlapping bulbs (or toes) metres across. Although lava in the tubes may rush down slope at several metres a second, pahoehoe fronts typically advance at less than 1 km a day.

phenocryst Magmatic rocks often contain at least two populations of crystals. One consists of tiny crystals that formed rapidly just before or during eruption. The second consists of much larger crystals that grew over long periods of time while the magma was underground. The large crystals are called phenocrysts.

phreatic eruption An eruption driven by non-volcanic water that has been vaporized to steam by the heat from ascending magma. The products are fragments of country rock alone. If new magma is also expelled, the eruption is hydromagmatic or phreatomagmatic.

Phreatic eruptions (from the Greek for a "well") commonly occur before the start of magmatic events.

phreatomagmatic See hydromagmatic.

pillow lava Collections of rounded and slightly flattened lava bodies resembling pillows up to metres across. Each pillow is a bulb of lava that has oozed from the margins of a flow as it travels under water. Similar structures are seen on land as collections of pahoehoe lava toes.

pisolite See lapilli.

Plinian eruption An eruption producing a buoyant cloud of ash and hot gas that may rise tens of kilometres into the atmosphere before spreading outwards. The column entrains cold surrounding air during ascent. Cooling increases the density of the cloud, especially its outer margins. If the cloud becomes too heavy, it collapses and crashes back to Earth (column collapse) as a pyroclastic flow that may still have temperatures exceeding 700°C and can race over the surface at hurricane velocities. The style of activity is named after Pliny the Younger, who described such behaviour during the AD 79 eruption of Somma.

porphyritic texture A texture describing a rock containing large crystals, or phenocrysts, within in a glassy or fine-grain crystalline matrix. Compare with aphyric texture.

pumice Highly vesicular magma normally produced during Plinian eruptions. The high vesicularity gives pumice its diagnostic density of 1000 kg per cubic metre or less, so that it can float on water (density of 1000 kg per cubic metre). Although normally associated with viscous magmas of intermediate and evolved compositions (e.g. andesite, dacite, trachyte, rhyolite and phonolite, and their potassic equivalents), pumice has been recorded from some basaltic eruptions. Pumiceous horizons can also form dur-

ing concentrated vesiculation in otherwise glassy lava flows (e.g. the Pietre Cotte flow on Vulcano).

pyroclastic Generic term describing volcanic rock broken during eruption (Greek: "fire-broken").

pyroclastic flow Cloud of hot gas and incandescent ash that, at temperatures of several hundred degrees Celsius, hugs the ground and races down slope at velocities of about 100 km per hour. Pyroclastic flows are commonly formed by the collapse of the cooler outer parts of an eruption column or by disintegration of a lava dome.

pyroclastic surge A pyroclastic flow consisting mostly of hot gas. Surges may occur as the dilute outer parts of a flow or they may be generated directly at the vent, especially during hydromagmatic eruptions when non-volcanic steam increases the gas available for expelling the magma.

quaquaversal From the rim of a volcanic cone, layers of lava and tephra are inclined in opposite directions along the cone's inner and outer flanks. Quaquaversal describes the change in inclination (forming an inverted V) when seen in outcrop.

rootless lava flow Lava flow produced by the mobilization of piles of fluid spatter, normally around lava fountains. The piles must accumulate rapidly for the whole mass to flow before individual pieces of spatter solidify.

scoria Generic term for broken volcanic rock with irregular rough surfaces and usually applied to describe aa fragments and the products, smaller than bombs, of Strombolian eruptions and lava fountains (Greek: "dung").

spatter Generic term for larger volcanic fragments (normally bombs) that have a rounded shape. They often become flattened during impact and are typically associated with Strombolian eruptions and lava fountains.

Strombolian eruption   The intermittent expulsion of blobs of magma up to metres across. This activity is typical of Stromboli and it occurs as large bubbles burst at the magma-filled throat of a volcano. The material ejected accumulates around the vent to produce new cones that may grow hundreds of metres tall.

tectonic plate   Earth's surface is made up of a few tectonic plates that jostle against each other under the influence of mantle movements. The plates are 70–120 km thick and consist of crust (continental or oceanic) and the rigid outer part of the mantle. Neighbouring plates can interact in four ways. At constructive plate margins, plates tear away from each other to allow underlying magma to reach the surface and form new crust; most such margins are found along the ocean floors. When two plates move towards each other, they may either collide and crumple to form mountain chains (collisional plate margin), or one plate may be forced to descend beneath the other (subduction zone). Parts of plates may also slide past each other at transform plate margins. Italy's volcanoes are a result of the collision between the African and Eurasian tectonic plates. In some places, this collision has forced subduction of the African plate (below the Aeolian Islands); elsewhere the jostling has caused local tearing along transform faults (e.g. the Roman Volcanic Province) and extensional fractures (e.g. Etna).

tephra   Generic term for all volcanic fragments that are explosively ejected (from the Greek for "ashes").

tuff   Consolidated pumice and ash deposit, usually associated with pyroclastic flows. Consolidation occurs as minerals deposited by circulating water cement the volcanic fragments.

vesicles   Gas bubbles within magma.

Vulcanian eruption   The sporadic expulsion of collections of bombs commonly metres across. They are driven by the accumulation of gas pressure beneath a thick cap of viscous magma that is blocking the volcanic conduit. The cap may have thicknesses of at least tens of metres, much greater than in the case for Strombolian eruptions. The resulting bombs often break into angular fragments. Their original surfaces are frequently dissected by a network of cracks, centimetres deep, that resemble the pattern on crusty bread (breadcrust bombs). The style of activity is named after the behaviour of Vulcano's 1888–90 eruption.

xenolith   Fragments of old rock trapped inside magma.

zeolitization   Fluids circulating through volcanic rock can deposit minerals along cracks and within vesicles. The minerals frequently belong to the zeolite family and can cement previously loose deposits.

Index

Printed and bound by CPI Group (UK) Ltd, Croydon, CR0 4YY

27/10/2024

14580407-0003